METHODS IN MOLECULAR BIOLOGY™

Series Editor
John M. Walker
School of Life Sciences
University of Hertfordshire
Hatfield, Hertfordshire, AL10 9AB, UK

For further volumes:
http://www.springer.com/series/7651

Epiblast Stem Cells

Methods and Protocols

Edited by

Ramiro Alberio

School of Biosciences, University of Nottingham, Loughborough, UK

Humana Press

Editor
Ramiro Alberio
School of Biosciences
University of Nottingham
Loughborough, UK

ISSN 1064-3745 ISSN 1940-6029 (electronic)
ISBN 978-1-62703-627-6 ISBN 978-1-62703-628-3 (eBook)
DOI 10.1007/978-1-62703-628-3
Springer New York Heidelberg Dordrecht London

Library of Congress Control Number: 2013945871

Printed on acid-free paper

Humana Press is a brand of Springer
Springer is part of Springer Science+Business Media (www.springer.com)

Preface

The cells of the epiblast give rise to all the tissues of the embryo proper. How epiblast cells acquire pluripotent features and how they transition to form the trilaminar embryo are current topics of intense investigation. A more detailed understanding of these in vivo events will lead to improved strategies for in vitro cellular differentiation protocols and genetic modification. In *Epiblast Stem Cells: Methods and Protocols*, leading researchers provide a detailed collection of techniques and protocols useful to the study of the biology of the pluripotent epiblast. As embryo development varies in model organisms, this volume includes chapters describing techniques used to study epiblast development in different amniotes. This collection brings contributions from the fields of embryology, stem cell biology and developmental biology together, such that readers can find in a single volume detailed procedures for the isolation and culture of epiblasts at different stages of development and techniques for the study of differentiation into specific lineages. Over the past five years, advances in the molecular characterization of pluripotent cells have led to a new understanding of how metastable states of pluripotency develop during the transition between the early inner cell mass to the late epiblast stage. Exploiting this new knowledge has enabled researchers to develop protocols that capture these novel pluripotent intermediates. These intermediate stages of pluripotency are important for modelling the transitions that these cells undergo in vivo and will contribute to improving the development of robust in vitro differentiation protocols.

Written in the highly successful *Methods in Molecular Biology*™ series format, chapters include introductions to their respective topics, a complete list of the necessary materials and reagents, detailed laboratory protocols, and extensive notes providing suggestions on troubleshooting and how to overcome common difficulties.

I hope the chapters presented in this volume will serve as an up-to-date source of novel protocols developed as a result of the new understanding in pluripotency and will contribute to the incremental use of stem cell technologies in different species.

Comprehensive and cutting-edge *Epiblast Stem Cells: Methods and Protocols* serves as a resource to individuals interested in studying the biology of pluripotent cells.

Loughborough, UK ***Ramiro Alberio***

Contents

Contributors

RAMIRO ALBERIO • *Division of Animal Sciences, School of Biosciences, University of Nottingham, Loughborough, UK*

CANTAS ALEV • *Laboratory for Early Embryogenesis, RIKEN Center for Developmental Biology, Chuo-Ku, Kobe, Japan*

PAULINE AUBEL • *Inserm, U846, Stem Cell and Brain Research Institute, Bron, France; Université de Lyon, Lyon, France*

SIQIN BAO • *Wellcome Trust/Cancer Research UK Gurdon Institute, University of Cambridge, Cambridge, UK*

ESMAIL BEHBOODI • *Department of Comparative Biology and Experimental Medicine, Faculty of Veterinary Medicine, University of Calgary, Calgary, AB, Canada*

DEAN H. BETTS • *Department of Physiology and Pharmacology, Schulich School of Medicine and Dentistry, University of Western Ontario, London, ON, Canada; Children's Health Research Institute, Lawson Health Research Institute, London, ON, Canada*

ALLA BONDAREVA • *Department of Comparative Biology and Experimental Medicine, Faculty of Veterinary Medicine, University of Calgary, Calgary, AB, Canada*

COURTNEY R. BROOKS • *Department of Physiology and Pharmacology, Schulich School of Medicine and Dentistry, University of Western Ontario, London, ON, Canada*

SHANBO CAO • *The Third Affiliated Hospital of Guangzhou Medical University, Guangzhou, China*

SILVIA COLLEONI • *Avantea, Laboratory of Reproductive Technologies, Cremona, Italy*

DAVID A. CONTRERAS • *Division of Animal Sciences, School of Biosciences, University of Nottingham, Loughborough, UK*

SEVERINE A. DEGRELLE • *INRA, UMR1198 Biologie du Développement et Reproduction, Jouy-en-Josas, France; ENVA, Maisons Alfort, France; Department of Immunology, INSERM, Institut Cochin, Paris, France*

SHENG DING • *Department of Pharmaceutical Chemistry, Gladstone Institute of Cardiovascular Disease, University of California, San Francisco, CA, USA*

INA DOBRINSKI • *Department of Comparative Biology and Experimental Medicine, Faculty of Veterinary Medicine, University of Calgary, Calgary, AB, Canada*

DANIEL C. FACTOR • *Department of Genetics and Genome Sciences, Case Western Reserve University School of Medicine, Cleveland, OH, USA*

CESARE GALLI • *Avantea, Laboratory of Reproductive Technologies, Cremona, Italy; Department of Veterinary Medical Sciences, University of Bologna, Bologna, Italy; Fondazione Avantea, Cremona, Italy*

WILLIAM G. GAVIN • *rEVO Biologics, Framingham, MA, USA*

ASTRID GILLICH • *Wellcome Trust/Cancer Research UK Gurdon Institute, University of Cambridge, Cambridge, UK*

VANESSA J. HALL • *Department of Veterinary Clinical and Animal Sciences, Faculty of Health and Medical Sciences, University of Copenhagen, Frederiksberg C, Denmark*

Katsuhiko Hayashi • *Department of Anatomy and Cell Biology, Graduate School of Medicine, Kyoto University, Kyoto, Japan*
Arata Honda • *OPTT, University of Miyazaki, Miyazaki, Japan; RIKEN BioResource Center, Ibaraki, Japan; Japan Science and Technology Agency, Saitama, Japan*
Hiroyuki Horiuchi • *Laboratory for Immunobiology, Department of Molecular and Applied Bioscience, Hiroshima University, Higashi-Hiroshima, Japan*
Isabelle Hue • *UMR Biologie du Développement et Reproduction, INRA, Jouy-en-Josas, France*
Poul Hyttel • *Department of Veterinary Clinical and Animal Sciences, Faculty of Health and Medical Sciences, University of Copenhagen, Frederiksberg C, Denmark*
Louise Lam • *Department of Comparative Biology and Experimental Medicine, Faculty of Veterinary Medicine, University of Calgary, Calgary, AB, Canada*
Giovanna Lazzari • *Avantea, Laboratory of Reproductive Technologies, Cremona, Italy; Fondazione Avantea, Cremona, Italy*
Wenlin Li • *Department of Cell Biology, Second Military Medical University, Shanghai, China*
Lin Liu • *College of Life Sciences, Nankai University, Tianjin, China*
Fadi J. Najm • *Department of Genetics and Genome Sciences, Case Western Reserve, University School of Medicine, Cleveland, OH, USA*
Mikiharu Nakano • *Laboratory for Immunobiology, Department of Molecular and Applied Bioscience, Hiroshima University, Higashi-Hiroshima, Japan*
Mark B. Nottle • *Reproductive Biotechnology Group, School of Paediatrics and Reproductive Health and Centre for Stem Cell Research, Robinson Institute, The University of Adelaide, Adelaide, SA, Australia*
Bertrand Pain • *U846 INSERM, USC7367 INRA, Stem Cell and Brain Research Institute, Bron, France; Université de Lyon, UMR S, Lyon, France*
Aida Rodriguez • *Division of Animal Sciences, School of Biosciences, University of Nottingham, Loughborough, UK*
Mikkel A. Rasmussen • *Department of Veterinary Clinical and Animal Sciences, Faculty of Health and Medical Sciences, University of Copenhagen, Frederiksberg C, Denmark*
Mitinori Saitou • *Department of Anatomy and Cell biology, Graduate School of Medicine, Kyoto University, Kyoto, Japan*
Guojun Sheng • *Laboratory for Early Embryogenesis, RIKEN Center for Developmental Biology, Chuo-Ku, Kobe, Hyogo, Japan*
M. Azim Surani • *Wellcome Trust/Cancer Research UK Gurdon Institute, University of Cambridge, Cambridge, UK*
Jonathan H. Teichroeb • *Department of Physiology and Pharmacology, Schulich School of Medicine and Dentistry, University of Western Ontario, London, ON, Canada; Children's Health Research Institute, Lawson Health Research Institute, London, ON, Canada*
Paul J. Tesar • *Department of Genetics and Genome Sciences, Case Western Reserve University School of Medicine, Cleveland, OH, USA; Department of Neurosciences, Case Western Reserve University School of Medicine, Cleveland, OH, USA*
Ian C. Tobias • *Department of Physiology and Pharmacology, Schulich School of Medicine and Dentistry, University of Western Ontario, London, ON, Canada*

IVAN VASSILIEV • *Reproductive Biotechnology Group, School of Paediatrics and Reproductive Health and Centre for Stem Cell Research, Robinson Institute, The University of Adelaide, Adelaide, SA, Australia*
CHRISTOPH VIEBAHN • *Department of Anatomy and Embryology, Centre of Anatomy, University of Göttingen, Göttingen, Germany*
FANG WANG • *State Key Laboratory of Medicinal Chemical Biology, College of Life Sciences, Nankai University, Tianjin, China; Department of Cell Biology and Genetics, College of Life Sciences, Nankai University, Tianjin, China*
YUPING WU • *Laboratory for Early Embryogenesis, RIKEN Center for Developmental Biology, Chuo-Ku, Kobe, Hyogo, Japan*

Chapter 1

Generation and Characterization of Epiblast Stem Cells from Blastocyst-Stage Mouse Embryos

Daniel C. Factor, Fadi J. Najm, and Paul J. Tesar

Abstract

Mouse epiblast stem cells (EpiSCs) are pluripotent embryonic cells that can be used to interrogate developmental transitions that occur during gastrulation. EpiSCs can also be robustly differentiated into functional somatic and germ cell derivatives making them a useful tool for studying development and regenerative medicine. Typically, mouse EpiSCs are isolated from the early postimplantation epiblast around 5.5 days post coitum (dpc). This chapter describes the methods for isolation of mouse EpiSCs from preimplantation blastocyst-stage mouse embryos (3.5 dpc). This technique enables the routine ability to derive EpiSC lines as it is much less labor intensive than isolation of EpiSCs from the postimplantation epiblast. We also detail relevant assays used to characterize new EpiSC lines and distinguish them from mouse embryonic stem cells.

Key words Epiblast, Pluripotency, Stem cell, Immunostaining, RNA-seq, Differentiation

1 Introduction

Pluripotent stem cells are widely utilized to study development and disease [1]. These cells can be expanded in the laboratory indefinitely while retaining the unique capacity to form any cell type of the body. Genetically normal pluripotent stem cells have been derived in vitro from multiple sources including (1) preimplantation embryos, (2) early postimplantation epiblast, (3) germ cells, and (4) mature somatic cells after exogenous modulation of gene expression (most typically by induction of Oct4, Sox2, Klf4, and c-myc) [2–12]. However, it is clear that not all pluripotent stem cell lines are equivalent. In fact, recent work has shown that irrespective of their origin, pluripotent stem cells can exist in one of the two developmental states referred to as "naïve" and "primed" [13].

Each pluripotent state provides unique properties that can be harnessed in the laboratory. Naïve pluripotent cells, typified by mouse embryonic stem cells (mESCs) and mouse induced pluripotent stem cells (iPSCs), have the advantage of being readily

Ramiro Alberio (ed.), *Epiblast Stem Cells: Methods and Protocols*, Methods in Molecular Biology, vol. 1074,
DOI 10.1007/978-1-62703-628-3_1, © Springer Science+Business Media New York 2013

amenable to genetic engineering by homologous recombination, contributing efficiently to all somatic and germ cell derivatives in chimeras, and being easily passaged by single-cell dissociation. Primed pluripotent cells such as mouse EpiSCs, human ESCs, and human iPSCs directly recapitulate developmental signaling processes that function during gastrulation. While naïve mouse pluripotent stem cells can be readily derived, isolation of primed mouse EpiSCs is a laborious process requiring isolation and manual dissection of the early postimplantation mouse embryo. We have recently shown that mouse EpiSCs can also be isolated from preimplantation blastocyst-stage embryos [12]. As blastocysts can be readily isolated, this technique should greatly enhance the ability of laboratories to derive their own mouse EpiSC lines in-house. This chapter provides detailed instructions for this procedure.

2 Materials

2.1 Preimplantation EpiSC Isolation

2.1.1 Harvesting the Blastocyst

1. Timed-pregnant female mouse at 3.5 days post coitum (dpc) (*see* **Note 1**).
2. Dissection scissors and forceps.
3. PBS: Phosphate-buffered saline without calcium and magnesium. PBS is stored at room temperature.
4. Dissection medium: FHM Hepes-buffered medium (Millipore). Dissection medium is stored in aliquots at −20 °C for up to 6 months. Once thawed, dissection medium can be used for up to 2 weeks.
5. 26-Gauge needle with 2 ml syringe (*see* **Note 2**).

2.1.2 Blastocyst Outgrowth

1. Nunc culture plates with the Nunclon Δ-treated surface (24-well).
2. Irradiated MEFs (iMEFs) can be purchased or derived in-house [14].
3. MEF culture medium: Dulbecco's modified Eagle medium supplemented with 10 % fetal bovine serum (FBS), 2 mM glutamax, 1× nonessential amino acids, and 0.1 mM 2-mercaptoethanol. Media components are combined and filtered through a 0.22 μm polyethersulfone filtration unit and stored at 4 °C for up to 4 weeks.
4. K15F5 culture medium: Knockout DMEM supplemented with 15 % knockout serum replacement (KSR), 5 % ES cell-qualified FBS, 2 mM glutamax, 1× nonessential amino acids, and 0.1 mM 2-mercaptoethanol. Media components are combined and filtered through a 0.22 μm polyethersulfone filtration unit and stored at 4 °C for up to 2 weeks.
5. Finely pulled glass pipette.
6. 0.25 % Trypsin–EDTA for blastocyst dissociation.

2.1.3 Picking EpiSCs

1. EpiSC culture medium: Knockout DMEM supplemented with 20 % KSR, 2 mM glutamax, 1× nonessential amino acids, and 0.1 mM 2-mercaptoethanol. Media components are combined and filtered through a 0.22 μm polyethersulfone filtration unit and stored at 4 °C for up to 2 weeks. Recombinant human FGF2 (10 ng/ml) and activin A (5 ng/ml) are freshly added to the media before use. FGF2 is dissolved in 0.1 % bovine serum albumin fraction V in PBS to a concentration of 10 ng/μl. FGF2 aliquots are stored at −80 °C and used within 6 months. Once thawed, FGF2 aliquots are stored at 4 °C and used within 2 weeks. Activin A is dissolved in 0.1 % bovine serum albumin fraction V in PBS to a concentration of 50 ng/μl, stored at −80 °C, and used within 6 months. Aliquots are thawed and diluted to 5 ng/μl in 0.1 % bovine serum albumin fraction V in PBS, stored at 4 °C, and used within 2 weeks.
2. Pulled glass pipette or 30-gauge needle.

2.2 Expansion of EpiSCs

2.2.1 Routine Culturc of EpiSCs

1. EpiSC passaging medium: 1.5 mg/ml collagenase type IV dissolved in EpiSC culture medium (without FGF2 or activin A) and filter sterilized with a 50 ml steriflip (Millipore). EpiSC passaging medium is stored at 4 °C and used within 2 weeks.
2. 96-Well round-bottom plate.

2.2.2 Freezing EpiSCs

1. 2× Freezing medium: 60 % Knockout DMEM, 20 % DMSO, and 20 % FBS. Filter sterilize the solution with a 0.22 μm syringe filter. Excess 2× freezing medium can be stored at 4 °C and used for up to 1 week.
2. 1.8 ml Cryotubes.
3. Cryo-freezing container: Nalgene "Mr. Frosty." Container is filled with 250 ml of isopropanol and changed after four uses. Store at room temperature between uses.

2.2.3 Thawing EpiSCs

1. 9.5 ml of EpiSC culture medium (without FGF2 or activin A) warmed in 37 °C water bath.
2. 5 ml of EpiSC culture medium with 10 ng/ml FGF2 and 5 ng/ml activin A, warmed in 37 °C water bath.

2.3 Characterization of EpiSCs

2.3.1 Immunostaining of Pluripotency Markers

1. 4 % Paraformaldehyde in PBS: In fume hood combine 10 ml of 16 % paraformaldehyde solution, 4 ml 10× PBS, and 26 ml of sterile water. Store in 4 °C for up to 10 days.
2. 0.2 % Triton-X in PBS: While stirring 500 ml PBS solution with a magnetic stir bar, slowly add 1 ml Triton-X solution. Stir gently for an extra 1 h. Store at room temperature. Solution is good for up to 1 year.
3. Blocking solution: Dilute 1 ml of donkey serum with 9 ml PBS solution and filter through a 0.22 μm Millex-GV filter attached to a Monoject luer lock syringe (Kendall). Store at 4 °C for up to 10 days.

4. Primary antibodies: Oct4 (Santa Cruz, SC-5279; 200 μg/ml) and Sox2 (R&D Systems, AF2018; 200 μg/ml). Oct4 antibody is stored at 4 °C and good for over a year. Sox2 aliquots are stored at −80 °C and good for 1 year. Once thawed, aliquots are kept for 1 month.
5. Secondary antibodies: Donkey anti-mouse 488 (Invitrogen, A21202) and donkey anti-goat 555 (Invitrogen, A21432).
6. DAPI stock solution: Add 1 ml water to 1 mg DAPI powder. Store at 4 °C and is good for over a year.

2.3.2 Global Gene Expression Analysis by RNA Sequencing

1. 1.7 ml RNase-free microcentrifuge tubes.
2. Trizol is aliquoted to prevent contamination and stored at 4 °C.
3. Phase lock gel tubes.
4. Chloroform.
5. Qiagen RNeasy Plus Micro Kit.
6. 70 % ethanol prepared with RNase-free water: Add 35 ml 200 proof ethanol to 15 ml nuclease-free water. Store at −20 °C so that solution is prechilled for RNA extraction procedure.
7. Illumina Truseq RNA Sample Preparation Kit.

2.3.3 Differentiation to Neuroectoderm

1. SB431542 is prepared in DMSO at a stock concentration of 20 mM. Aliquots are stored at −80 °C and good for 6 months. Tubes are freeze-thawed a maximum of three times.
2. Noggin is prepared in 0.1 % bovine serum albumin fraction V in PBS at 100 ng/μl. Aliquots are stored at −80 °C and good for 6 months. Tubes are thawed and stored at 4 °C and used within 2 weeks.
3. Dorsomorphin is prepared in DMSO solution at 10 mM. Aliquots are stored at 4 °C and good for 6 months.
4. Neural culture media: DMEM/F12 with 1× N2 Max supplement, 1× B-27 without vitamin A, and 2 mM glutamax. Media components are combined and filtered through a 0.22 μm polyethersulfone filtration unit and stored at 4 °C for up to 2 weeks.

3 Methods

3.1 Isolation and Maintenance of Preimplantation Epiblast

3.1.1 Isolation of 3.5 dpc Preimplantation Mouse Embryos

1. Euthanize female mouse at 3.5 dpc according to local animal care requirements.
2. Cut open abdominal cavity, dissect out the uterus, and transfer to a 100 mm petri dish filled with PBS. Cut away excess fat and connective tissue from the uterus.
3. Transfer the uterus to a 35 mm petri dish filled with dissection medium.
4. With a 26-gauge needle flush out each horn toward the cervix to release blastocysts into the culture dish.

3.1.2 Blastocyst Outgrowth

1. Using a stereomicroscope collect blastocysts and plate each into an individual 24-well containing iMEFs prepared the previous day (*see* **Note 3**). Primary blastocysts are grown in K15F5 culture medium.
2. Incubate blastocysts in standard tissue culture incubator for 5–6 days for the blastocyst outgrowth to form. Do not disturb or change the medium.
3. With a finely pulled glass pipette, dislodge the blastocyst outgrowth off the plate and place into 10 μl of 0.25 % trypsin–EDTA for 5 min in an individual well of a round-bottom 96-well plate.
4. Partially dissociate the blastocyst with a 10 μl pipette, being sure not to produce a single-cell suspension.
5. In K15F5 culture medium, plate each partial dissociate into an individual 24-well containing iMEFs prepared the previous day.
6. Incubate in tissue culture incubator for 6 days. Medium can be changed on the third day.
7. Passage each well on day 6: Rinse each 24-well with PBS (*see* **Note 4**) and add 250 μl 0.25 % trypsin–EDTA for 2–3 min.
8. Inactivate trypsin–EDTA with 250 μl of MEF culture media and gently triturate the cells being careful not to produce a single-cell suspension.
9. Transfer each well of cells to an individual 6-well with 2.5 ml EpiSC culture medium containing iMEFs prepared the previous day.
10. Incubate in tissue culture incubator for 4–8 days. During this time morphologically distinct mESC and EpiSC colonies become evident (Fig. 1a). A representative mESC colony can be seen in Fig. 1b and a representative EpiSC colony in Fig. 1c.

3.1.3 Picking EpiSCs

1. Rinse a 24-well containing iMEFs prepared the previous day with PBS. Add 1 ml EpiSC culture medium and place back into the incubator to equilibrate.
2. Under a stereomicroscope in a sterile hood, use a glass pipette or a 30-gauge needle to score individual EpiSC colonies into 4–6 equally sized pieces.
3. Transfer the pieces to the 24-well with EpiSC culture medium. Culture for 2–3 days changing the medium daily.
4. Under optimal conditions, you can expect 25–40 % of blastocysts to yield an EpiSC line.

3.2 EpiSC Culture

3.2.1 Routine Passaging of EpiSCs

1. After the first passage, EpiSCs can be passaged enzymatically. To passage, remove EpiSC culture medium from each well, add EpiSC passaging medium (250 μl for a 24-well or 1 ml for a 6-well), and incubate at 37 °C for 8–12 min.

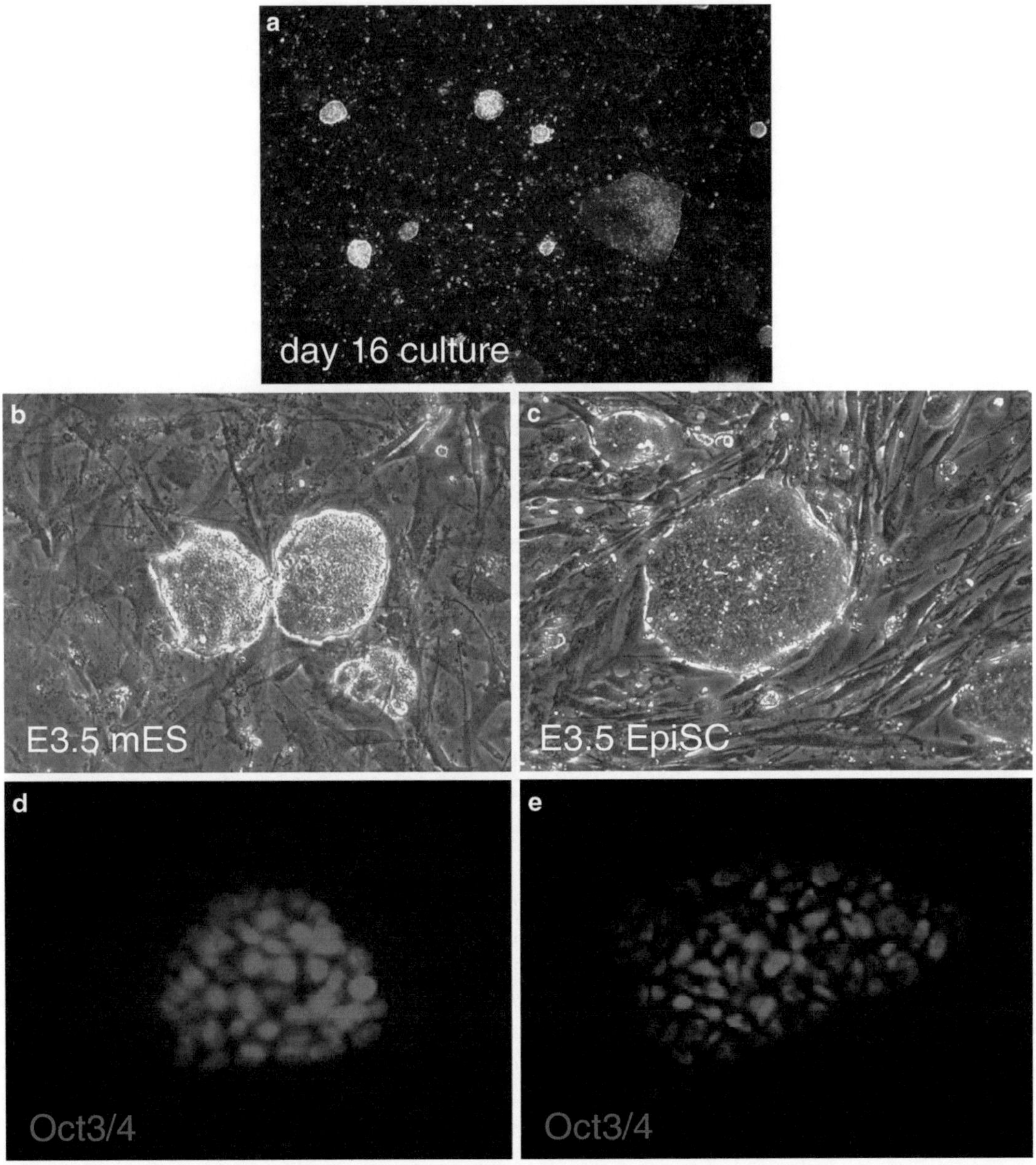

Fig. 1 Distinct pluripotent stem cell states can be derived from cultured mouse blastocysts. (**a**) After 16 days of culture, characteristic flat (EpiSC-like) or domed (mESC-like) colonies begin to emerge. (**b**) Morphology of mouse embryonic stem cell colonies with the characteristic domed shape. (**c**) Morphology of a blastocyst-derived epiblast stem cell colony with the flat, epithelial characteristics. Representative immunostaining of pluripotency marker Oct4 in an embryonic stem cell colony (**d**) and a preimplantation epiblast stem cell colony (**e**)

2. Add equivalent volume of EpiSC culture medium to each well and dislodge colonies away from iMEFs by gentle pipetting with a 5-ml pipette. Combine colony suspensions from each plate into a 15-ml conical tube.

3. Separate colonies away from any remaining MEFs by centrifugation at 200 × *g* for 30–40 s. The colonies will loosely pellet at the bottom of the tube while the individual MEFs will remain in the supernatant.
4. Aspirate the supernatant and discard. Gently resuspend the colony pellet in 5 ml EpiSC culture medium (FGF2 and activin are omitted from EpiSC culture medium when used for rinsing) and perform another 30–40-s spin. Repeat rinse/spin cycle one more time for a total of three.
5. Resuspend colonies in 100 μl of EpiSC culture medium and transfer to an individual well of a 96-well round-bottom plate.
6. Triturate colonies with a P100 pipette set at 80 ml until only small clusters remain (*see* **Note 5**).
7. Add triturated suspension to pre-warmed EpiSC culture medium and plate 2.5 ml per 6-well with iMEFs. EpiSCs are typically split 1:4–1:6 every 2–3 days. The density at which each EpiSC line is grown is very important and must be determined empirically. We continuously culture EpiSCs in 6-wells.
8. EpiSC culture medium is changed daily.

3.2.2 Cryopreservation of EpiSCs

1. Isolate colonies of EpiSCs free from MEFs as described in Subheading 3.2.1.
2. Triturate into small clusters just slightly larger than for passaging and spin down at 200 × *g* for 4 min.
3. Resuspend pellet in MEF culture medium at a concentration of 0.5 ml for every 35 mm well being frozen. Add an equal volume of 2× freezing medium, mix gently by inverting the tube, and distribute 0.5 ml of cell suspension into individual cryotubes (final volume of 0.5 well/vial). Place tubes in a room-temperature cryo-freezing container and immediately place at −80 °C. The following day, transfer vials to a liquid nitrogen freezer for permanent storage.

3.2.3 Thawing of EpiSCs

1. Remove vial from liquid nitrogen storage and thaw in 37 °C water bath until only a small ice crystal remains. Rinse vial with 1 ml of MEF culture medium and combine with 8.5 ml MEF culture medium in 15 ml conical (total of 10 ml). Spin down at 200 × *g* for 4 min, aspirate supernatant, and gently resuspend pellet in 5 ml of pre-warmed EpiSC culture medium. Remove MEF culture medium from iMEFs that were seeded the prior day, rinse with PBS, and aliquot the suspension equally into two 35 mm wells. Place in tissue culture incubator.
2. If recovery is adequate, colonies are ready to passage in 2–3 days.

3.3 Characterization of EpiSCs

3.3.1 Immunostaining of Pluripotency Markers

1. Aspirate cell culture medium from the cells and add 4 % paraformaldehyde in PBS to fix cells. Store the plate on a rocker for 15 min.
2. Aspirate the fixative and rinse 3× with PBS. Aspirate PBS, and then add 0.2 % Triton-X in PBS for 5 min while rocking. Repeat the Triton-X step for another 5 min.
3. Aspirate Triton-X solution, and then add 10 % donkey serum diluted in PBS for 2 h rocking at room temperature (*see* **Note 6**).
4. Add primary antibodies Oct4 (1:500) and Sox2 (1:100) diluted in blocking solution for 1 h rocking at room temperature or overnight at 4 °C.
5. Aspirate antibody solution, wash once with PBS, then add PBS, and incubate for 5 min while rocking. Repeat this step two more times.
6. Add secondary antibodies donkey anti-mouse (1:500) and donkey anti-goat (1:500) diluted in blocking solution for 1 hr rocking at room temperature. Keep the plate protected from light from this point onward by wrapping in aluminum foil.
7. Aspirate, wash once with PBS, then add PBS, and incubate for 5 min while rocking. Repeat this step two more times.
8. Incubate in 1 μg/ml DAPI solution in PBS (dilute DAPI stock solution 1:1,000) for 5 min with rocking.
9. Rinse with PBS and incubate for 5 min. Repeat this step. Never let the wells dry out.
10. Cells are ready for visualization. Plates can be stored at 4 °C for over 3 months as long as they are sealed with parafilm and covered in foil. EpiSC cell lines will be uniformly positive for Oct4 (Fig. 1d) and Sox2 with only rare spontaneously differentiated colonies being negative for these markers (*see* **Note 7**).

3.3.2 Global Gene Expression Analysis by RNA Sequencing (See Note 8)

1. Harvest EpiSCs from MEFs as described in Subheading 3.2.1, **steps 1–4**.
2. Transfer ~1×10^6 cells to a 1.5 ml RNase-free microcentrifuge tube and centrifuge the sample at $200 \times g$ for 4 min.
3. Carefully remove the supernatant and resuspend the sample in 0.5 ml Trizol. Samples may be processed immediately, or placed at −80 °C for long-term storage.
4. Prepare a phase lock gel tube for each sample, by spinning down at maximum speed for 3 min.
5. Add 0.5 ml Trizol to each sample (to bring the total volume to 1 ml) and transfer to the phase lock tubes.
6. Add 0.2 ml chloroform to each sample and shake vigorously for 15–20 s.

7. Centrifuge phase lock tubes at 4,000 × *g* for 3 min at room temperature to separate.
8. Collect the aqueous phase from each sample and transfer to a gDNA elimination column (supplied with Qiagen RNeasy Plus Micro Kit).
9. Centrifuge the column at 8,000 × *g* for 30 s. If liquid remains on the membrane, repeat the spin until none remains. Save the flow through, and discard the column.
10. Add 1 volume of cold 70 % ethanol prepared with RNase-free water to each sample, and pipet up and down until homogeneous.
11. Transfer no more than 700 μl at a time to RNeasy MinElute columns (supplied with Qiagen RNeasy Plus Micro Kit).
12. Centrifuge at 8,000 × *g* for 15 s. Discard the flow through, and repeat for any remaining volume.
13. Proceed with RNeasy Plus Micro "Purification of total RNA from Animal and Human Cells" protocol according to the manufacturer's instructions, beginning with **step** 7 (07/2007 protocol version).
14. Check the concentration of RNA immediately following purification using a Nanodrop spectrophotometer. Samples should be used immediately or stored at −80 °C until use.
15. Verify quality and integrity of RNA samples using a Bioanalyzer or a similar instrument immediately prior to proceeding with library preparation. An RNA integrity number (RIN) or RNA quality indicator (RQI) >8 and 1 μg of RNA provide a good starting point for library preparation.
16. Carry out mRNA enrichment by two rounds of poly-A selection, as well as RNA fragmentation, cDNA synthesis, and sequencing library preparation using the Illumina Truseq RNA Sample Preparation Kit according to the manufacturer's instructions.
17. Sequence samples on an Illumina instrument (HiSeq or similar). We find that >40 million 100 bp paired-end reads provide sufficient sequencing depth for confidently interrogating the expression level of known genes (*see* **Note 9**).
18. Align RNA seq reads to the mm9 reference genome (*see* **Note 10**) using the Tophat spliced-read aligner [15, 16]. Use default settings unless reads are shorter than 50 base pairs, in which case the --segment-length option should be changed to half of the read length.
19. Estimate gene and transcript expression levels using Cufflinks (if only a single sample is being examined) or Cuffdiff (if multiple conditions are being compared, or if multiple replicates are available) software packages [17, 18]. Limit both of these programs to known transcripts by providing the mm9 RefSeq gene

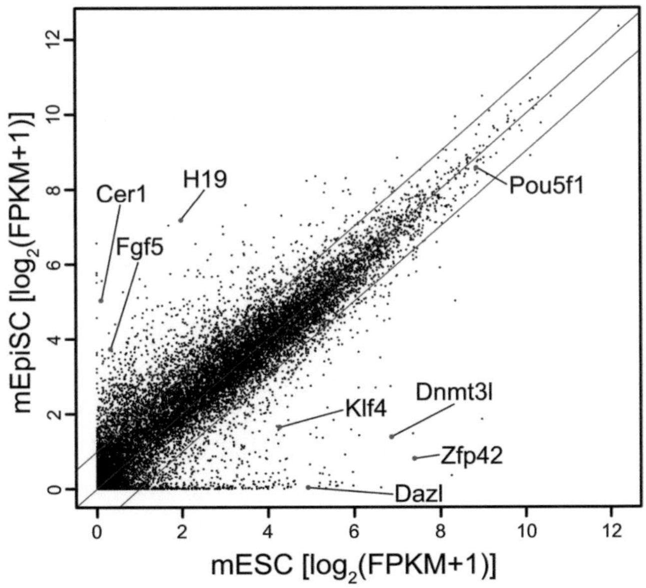

Fig. 2 Distinct gene expression signatures of EpiSCs and ESCs as revealed by RNA sequencing. EpiSCs and ESCs display distinctive expression patterns. Expression values are expressed as fragments per kilobase exon per million reads, or FPKM. One is added to the FPKM value to allow logarithm to be calculated. While key pluripotency regulatory factor Pou5f1 (also known as Oct3/4) is expressed at a similar level in the two cell types, Klf4 expression is known to be stronger in mouse ESCs. Examples of other genes known to distinguish these cell types are also highlighted

annotation GTF file (*see* **Note 10**) using the --GTF option (NOT --GTF-guide). Provide the whole-genome sequence using the --frag-bias-correct option to correct for sequence biases which may be introduced during the library preparation.

20. Both Cufflinks and Cuffdiff output a genes.fpkm_tracking file containing the expression values of all RefSeq genes, and an isoforms.fpkm_tracking file containing expression values of each individual transcript. View the contents of these files using standard spreadsheet or text-editing software. EpiSCs have a distinct gene expression signature as compared to mESCs (Fig. 2) [9, 19].

3.3.3 Differentiation to Neuroectoderm

1. Harvest and prepare EpiSC colonies for seeding as described in Subheading 3.2.1.
2. Seed cells 1:4–1:6 into wells of iMEFs in EpiSC culture medium supplemented with 20 μM SB431542 to inhibit activin/Nodal signaling. 100 ng/ml noggin and 2 μM dorsomorphin may also be added to inhibit BMP signaling (*see* **Note 11**).

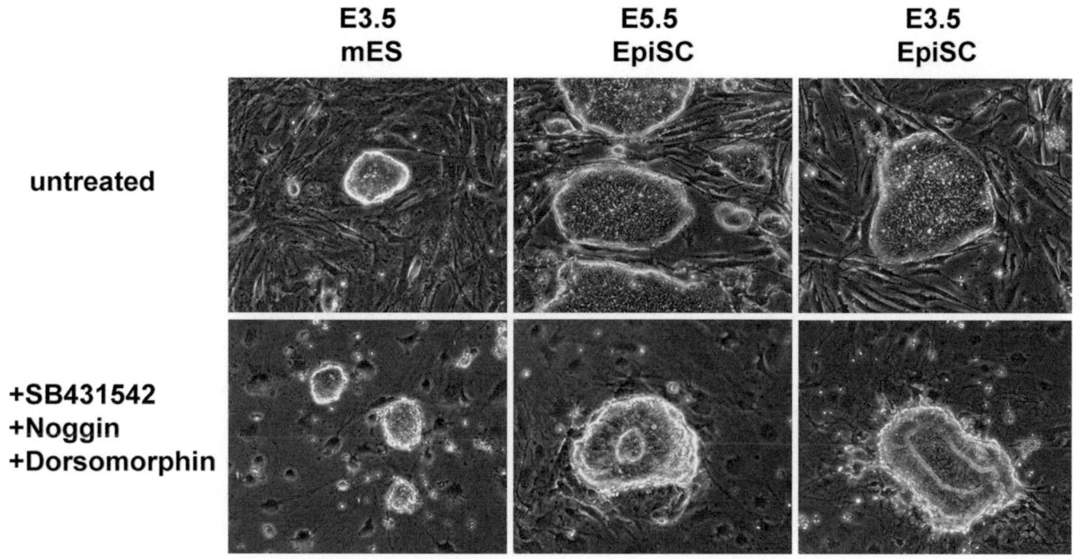

Fig. 3 Generation of neuroectoderm from EpiSCs. Epiblast stem cells derived from preimplantation and post-implantation embryos differentiate into neuroectoderm when treated with SB431542, noggin, and dorsomorphin. Embryonic stem cells do not form neuroectoderm under this scheme

3. Incubate cells in tissue culture incubator for 1 day.
4. Change cell media to 50 % EpiSC culture medium and 50 % neural culture media supplemented with 20 μM SB431542 (100 ng/ml noggin and 2 μM dorsomorphin may be included).
5. Incubate cells in tissue culture incubator for 1 day.
6. Change cell media to 100 % neural culture media supplemented with 20 μM SB431542 (100 ng/ml noggin and 2 μM dorsomorphin may be included).
7. Incubate cells in tissue culture incubator for 1 day.
8. Change cell media to 100 % neural culture media supplemented with 100 ng/ml noggin.
9. Incubate cells in tissue culture incubator for 1 day.
10. Change cell media to 100 % neural culture media supplemented with 100 ng/ml noggin.
11. At this point neural rosettes will have formed in nearly all of the colonies; *see* Fig. 3 for characteristic shape. Rosettes will be positive for both Pax6 and Sox1 markers. Cells can be differentiated further to a specific downstream neural lineage or dissociated and maintained as progenitors in the presence of FGF [20].

4 Notes

1. The mouse strain used for EpiSC derivation is not known to influence derivation efficiency.
2. Traditionally mouth pipettes have been used to transfer preimplantation embryos. This provides precise control although local safety requirements may not allow for the use of mouth pipettes. *See* Downs [21] for details.
3. When plating iMEFs, tissue culture plates must be pretreated with sterile 0.1 % gelatin Type A from porcine skin for at least 30 min prior to seeding cells. iMEFs are plated at a density of 5.0×10^4 cells/cm^2.
4. Rinsing with PBS is done to remove traces of serum from the cells. This is particularly important when switching to EpiSC culture medium as it does not contain serum.
5. It is important to dissociate the colonies in a well of a 96-well round-bottom plate so that you can monitor your dissociation. We typically triturate colonies ten times and check their size under the microscope. EpiSCs do not survive well as single cells, so it is important to reach an optimal size and not triturate too much.
6. We block with donkey serum because we typically buy secondary antibodies derived from donkey. If using secondary antibody derived from goat then block with goat serum.
7. Troublesome EpiSC cultures may show extensive differentiation and heterogeneous Oct4 expression. Manually selecting morphologically undifferentiated colonies and expanding them individually can typically rescue these cultures.
8. We recommend using a designated workspace and set of reagents for all work involving RNA. This reduces the risk of contamination and simplifies troubleshooting of RNA protocols.
9. It is important to produce reads of at least 50 base-pair length, as reads are split into multiple segments for alignment, and 25 base pairs is the recommended segment length for the Tophat aligner.
10. Illumina has compiled reference genomes compatible with Tophat and Cufflinks that contain the files required for this protocol, including the bowtie indices, RefSeq gene annotation GTF file, and whole-genome fasta file. These are available for download at http://tophat.cbcb.umd.edu/igenomes.html.
11. SB431542 and dorsomorphin can precipitate in the DMSO stock solutions. Therefore it is very important to heat each tube up to 50 °C to ensure that the small molecules are fully in solution prior to adding them to culture media.

References

1. Chenoweth JG, McKay RD, Tesar PJ (2010) Epiblast stem cells contribute new insight into pluripotency and gastrulation. Dev Growth Differ 52:293–301
2. Matsui Y, Zsebo K, Hogan BL (1992) Derivation of pluripotential embryonic stem cells from murine primordial germ cells in culture. Cell 70:841–847
3. Resnick JL, Bixler LS, Cheng L, Donovan PJ (1992) Long-term proliferation of mouse primordial germ cells in culture. Nature 359: 550–551
4. Thomson JA, Itskovitz-Eldor J, Shapiro SS et al (1998) Embryonic stem cell lines derived from human blastocysts. Science 282: 1145–1147
5. Evans MJ, Kaufman MH (1981) Establishment in culture of pluripotential cells from mouse embryos. Nature 292:154–156
6. Martin GR (1981) Isolation of a pluripotent cell line from early mouse embryos cultured in medium conditioned by teratocarcinoma stem cells. Proc Natl Acad Sci U S A 78:7634–7638
7. Brook FA, Gardner RL (1997) The origin and efficient derivation of embryonic stem cells in the mouse. Proc Natl Acad Sci U S A 94: 5709–5712
8. Tesar PJ (2005) Derivation of germ-line-competent embryonic stem cell lines from preblastocyst mouse embryos. Proc Natl Acad Sci U S A 102:8239–8244
9. Tesar PJ, Chenoweth JG, Brook FA et al (2007) New cell lines from mouse epiblast share defining features with human embryonic stem cells. Nature 448:196–199
10. Brons IG, Smithers LE, Trotter MW et al (2007) Derivation of pluripotent epiblast stem cells from mammalian embryos. Nature 448: 191–195
11. Takahashi K, Yamanaka S (2006) Induction of pluripotent stem cells from mouse embryonic and adult fibroblast cultures by defined factors. Cell 126:663–676
12. Najm FJ, Chenoweth JG, Anderson PD et al (2011) Isolation of epiblast stem cells from preimplantation mouse embryos. Cell Stem Cell 8:318–325
13. Nichols J, Smith A (2009) Naive and primed pluripotent states. Cell Stem Cell 4:487–492
14. Chenoweth JG, Tesar PJ (2010) Isolation and maintenance of mouse epiblast stem cells. Methods Mol Biol 636:25–44
15. Trapnell C, Pachter L, Salzberg SL (2009) TopHat: discovering splice junctions with RNA-Seq. Bioinformatics 25:1105–1111
16. Langmead B, Trapnell C, Pop M, Salzberg SL (2009) Ultrafast and memory-efficient alignment of short DNA sequences to the human genome. Genome Biol 10:R25
17. Roberts A, Trapnell C, Donaghey J et al (2011) Improving RNA-Seq expression estimates by correcting for fragment bias. Genome Biol 12:R22
18. Trapnell C, Williams BA, Pertea G et al (2010) Transcript assembly and quantification by RNA-Seq reveals unannotated transcripts and isoform switching during cell differentiation. Nat Biotechnol 28:511–515
19. Jiang J, Chan YS, Loh YH et al (2008) A core Klf circuitry regulates self-renewal of embryonic stem cells. Nat Cell Biol 10:353–360
20. Najm FJ, Zaremba A, Caprariello AV et al (2011) Rapid and robust generation of functional oligodendrocyte progenitor cells from epiblast stem cells. Nat Methods 8:957–962
21. Downs KM (2006) In vitro methods for studying vascularization of the murine allantois and allantoic union with the chorion. Methods Mol Med 121:241–272

Chapter 2

Reversion of Mouse Postimplantation Epiblast Stem Cells to a Naïve Pluripotent State by Modulation of Signalling Pathways

Astrid Gillich, Siqin Bao, and M. Azim Surani

Abstract

Mouse postimplantation epiblast cultured in activin and basic fibroblast growth factor gives rise to continuously growing epiblast stem cells (EpiSCs) that share key properties with postimplantation epiblast, such as DNA methylation and an inactive X-chromosome. EpiSCs also show a distinct gene expression profile compared to embryonic stem cells (ESCs) derived from preimplantation blastocysts, and do not contribute efficiently to chimeras. EpiSCs can, however, revert to pluripotent ESC-like cells upon exposure to leukemia inhibitory factor–Stat3 signalling on feeder cells. Here we describe a protocol for the establishment of EpiSCs and their reversion to ESCs.

Key words Epiblast stem cells, Embryonic stem cells, Postimplantation epiblast, Naïve pluripotency, Epigenetic reprogramming, LIF–Stat3 signalling

1 Introduction

The transition from preimplantation to postimplantation development is accompanied by extensive morphological and molecular changes of the rodent embryo [1]. All somatic cell lineages as well as the germ cells originate from the epiblast that transforms into a cup-shaped epithelium, acquires DNA methylation, and undergoes X-chromosome inactivation in female embryos [1–3]. Mouse postimplantation epiblast cultured in activin A and basic fibroblast growth factor (bFGF) gives rise to self-renewing cell lines termed epiblast stem cells (EpiSCs) [4, 5]. EpiSCs express the core pluripotency factors Oct4, Sox2, and Nanog and show multi-lineage differentiation potential in teratoma assays, but unlike embryonic stem cells (ESCs) derived from preimplantation blastocysts, they do not contribute efficiently, if at all, to chimeras [5, 6]. EpiSCs share key properties with postimplantation epiblast including the use of the proximal enhancer to drive *Oct4* expression instead of

Ramiro Alberio (ed.), *Epiblast Stem Cells: Methods and Protocols*, Methods in Molecular Biology, vol. 1074,
DOI 10.1007/978-1-62703-628-3_2, © Springer Science+Business Media New York 2013

the distal enhancer as in ESCs and primordial germ cells [6–8]. In addition, some pluripotency and germ cell-associated genes such as *Stella* and *Rex1* are methylated in EpiSCs [7].

We have previously shown that EpiSCs can revert to pluripotent ESC-like cells upon exposure to leukemia inhibitory factor (LIF)–Stat3 signalling on feeder cells [7]. Loss of DNA methylation, activation of the distal *Oct4* enhancer, X-chromosome reactivation, and re-expression of *Stella* and *Rex1* occur during the reversion process [7]. Here we describe a protocol for reversion of EpiSCs to ESC-like cells, termed reverted ESCs (rESCs). We first discuss the establishment of primary mouse embryonic fibroblasts (MEFs) for the preparation of feeder cells. We then explain how to derive EpiSCs from embryonic day (E) 5.5–6.5 postimplantation epiblast and we finally provide a detailed protocol for the reversion of EpiSCs to rESCs.

2 Materials

2.1 Cell Culture Media

1. Sterile Dulbecco's modified Eagle medium/Ham's Nutrient Mixture F-12 (DMEM/F-12, 1:1, Gibco).
2. Sterile MEF medium: DMEM supplemented with 10 % (v/v) fetal calf serum (FCS), 2 mM L-glutamine, 100 units/mL penicillin, and 100 μg/mL streptomycin. Mix 440 mL of DMEM with 50 mL of FCS, 5 mL of L-glutamine (200 mM stock solution), and 5 mL of penicillin/streptomycin solution (10,000 units/mL and 10,000 μg/mL stock solution). Store at 4 °C for up to 2 weeks.
3. Sterile EpiSC medium: Ham's Nutrient Mixture F-12 with GlutaMAX/Iscove's modified Dulbecco's medium (IMDM) (1:1) supplemented with 0.5 % (w/v) bovine serum albumin (BSA), 450 μM monothioglycerol, 7 μg/mL insulin, 15 μg/mL transferrin, 1× chemically defined lipid concentrate, 100 units/mL penicillin, 100 μg/mL streptomycin, 20 % (v/v) knockout serum replacement, 20 ng/mL activin A, and 12 ng/mL bFGF. Dissolve 2.5 g of BSA (Europa Bioproducts) in 250 mL of Ham's Nutrient Mixture F-12 with GlutaMAX, 250 mL of IMDM, 20 μL of monothioglycerol (≥97 %), 140 μL of human insulin (25 mg/mL stock solution), 75 μL of bovine apo-transferrin (100 mg/mL stock solution), 5 mL of chemically defined lipid concentrate (100× stock solution), and 5 mL of penicillin/streptomycin solution (10,000 units/mL and 10,000 μg/mL stock solution). Filter through a 0.45 μm membrane and store at 4 °C for up to 2 weeks. Before use, add 2 mL of knockout serum replacement, 4 μL of human activin A (50 μg/mL stock solution), and 2.4 μL of bFGF (50 μg/mL stock solution) to 8 mL of medium.

4. Sterile ESC medium: DMEM/F-12 (1:1) supplemented with 2 mM L-glutamine, nucleosides (30 μM adenosine, 30 μM guanosine, 30 μM cytidine, 30 μM uridine, 10 μM thymidine), 0.1 mM MEM nonessential amino acids (NEAA), 1 mM sodium pyruvate, 1.2 mg/mL sodium bicarbonate, 0.05 mM 2-mercaptoethanol, 100 units/mL penicillin, 100 μg/mL streptomycin, 20 % (v/v) FCS, and 1,000 units/mL LIF. Combine 367 mL of DMEM/F-12 (1:1) with 5 mL of L-GLUTAMINE (200 mM stock solution), 5 mL of nucleosides (100× stock solution, *see* **Note 1**), 5 mL of MEM NEAA (10 mM stock solution), 5 mL of sodium pyruvate (100 mM stock solution), 8 mL of sodium bicarbonate (7.5 % stock solution), 500 μL of 2-mercaptoethanol (50 mM stock solution), and 5 mL of penicillin/streptomycin solution (10,000 units/mL and 10,000 μg/mL stock solution). Store at 4 °C for up to 4 weeks. Before use, add 2 mL of FCS (*see* **Note 2**) and 1 μL of LIF (10,000,000 units/mL stock solution, ESGRO, Chemicon) to 8 mL of medium. Upon addition of FCS and LIF, store medium at 4 °C for up to 1 week.

2.2 Buffers and Solutions

1. Sterile phosphate-buffered saline (PBS) pH 7.4, calcium/magnesium-free.
2. Sterile distilled water (dH_2O).
3. Sterile 0.1 % (w/v) gelatin solution: Dissolve 0.1 g gelatin in 100 mL dH_2O in a 55 °C water bath. Autoclave and store at room temperature for up to 4 weeks.
4. Sterile collagenase solution: Knockout DMEM supplemented with 10 % (v/v) knockout serum replacement, 0.1 % (w/v) collagenase, 2 mM L-glutamine, 0.1 mM MEM NEAA, and 0.35 μM 2-mercaptoethanol. Dissolve 500 mg of Collagenase Type IV (Invitrogen) in 400 mL of Knockout DMEM (Gibco) supplemented with 100 mL of knockout serum replacement (Gibco), 5 mL of L-glutamine (200 mM stock solution), 5 mL of MEM NEAA (10 mM stock solution), and 3.5 μL of 2-mercaptoethanol (50 mM stock solution). Filter through a 0.45 μM membrane and store at 4 °C for up to 1 week.
5. Sterile 0.05 % (w/v) trypsin/EDTA (1×) solution.
6. Sterile trypsin/pancreatin solution: 0.5 % (w/v) trypsin and 2.5 % (w/v) pancreatin in calcium/magnesium-free Tyrode Ringer's Saline pH 7.6–7.7. Dissolve 0.1 g trypsin and 0.5 g pancreatin in 20 mL of Tyrode Ringer's Saline (*see* **Note 3**). Filter through a 0.45 μM membrane and store 1 mL aliquots at –20 °C.
7. Sterile mitomycin C solution (1 mg/mL in PBS): Store at 4 °C protected from light for up to 1 week or at –80 °C for up to 1 year.
8. Sterile dimethylsulfoxide (DMSO).

2.3 Equipment

1. Two pairs of forceps (Dumont #5).
2. Fine scissors.
3. Surgical blade.
4. Borosilicate glass capillaries (1.0 mm outside diameter, 0.58 mm inside diameter, Sutter Instruments).
5. Micropipette puller (Sutter Instruments).
6. Mouth pipette (*see* **Note 4**).
7. Stereomicroscope.
8. Sterile 90 mm bacteriological dishes.
9. Sterile 19 G needles.
10. Sterile 5 mL syringes.
11. Sterile 0.45 μM filters.
12. Sterile 90 mm tissue culture dishes.
13. Sterile 24-well (1.9 cm^2), 12-well (3.5 cm^2), 6-well (9.6 cm^2), and 4-well (1.9 cm^2) multidishes.
14. Sterile 30 mL polypropylene tubes.
15. Sterile 25, 10, and 5 mL pipettes.
16. Sterile cryovials.
17. Cryo container.
18. Benchtop centrifuge.
19. Cell culture incubator: Sterile, humidified, 37 °C, 5 % CO_2, 95 % air.
20. Tissue culture hood.
21. Liquid nitrogen tank and −80 °C freezer.
22. Water bath.
23. Hot plate.
24. Vacuum pump.
25. Fluorescence microscope.

2.4 Animals

1. Pregnant female mice sacrificed humanely at 13.5 days of gestation for MEF derivation.
2. Pregnant female mice sacrificed humanely at 5.5–6.5 days of gestation for EpiSC derivation (preferably with *Oct4*-ΔPE-GFP reporter, *see* **Note 5**) [7, 8].

3 Methods

3.1 Establishment of Primary MEFs for Preparation of Feeder Cells

1. Prepare sterile MEF medium.
2. Open the abdominal cavity and remove the uterus from one to two pregnant female mice at 13.5 days of gestation.

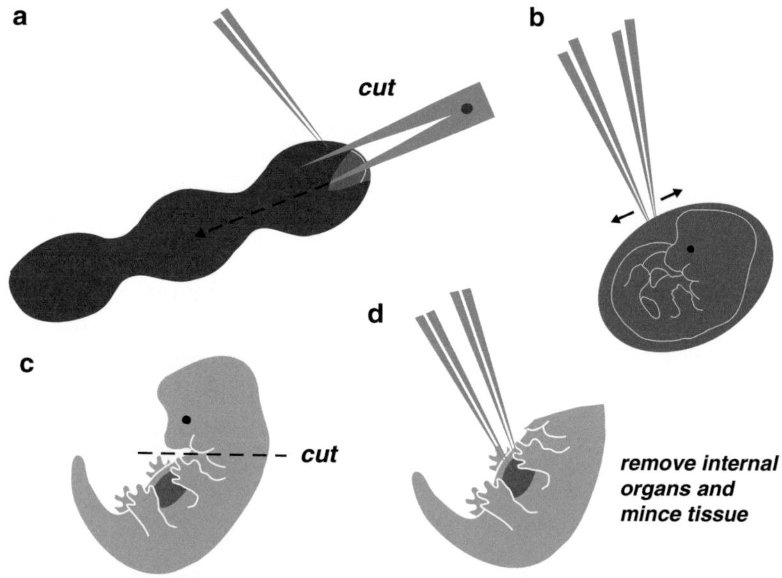

Fig. 1 Derivation of primary MEFs from E13.5 embryos. The uterine wall is cut (**a**) and the extraembryonic membranes are removed (**b**). Following removal of the head and the internal organs (**c**, **d**), the tissue is minced, transferred to a dish containing sterile MEF medium, and incubated in a humidified incubator at 37 °C and 5 % CO_2

3. Place the uterus into a 90 mm bacteriological dish containing 10 mL of sterile PBS and cut the uterine wall to peel out the deciduae (Fig. 1a).
4. Use two pairs of forceps to cut the Reichert's membrane and the placenta (Fig. 1b). Remove the extraembryonic membranes and transfer the embryos to a new 90 mm bacteriological dish containing 10 mL of sterile PBS.
5. Cut off the head (Fig. 1c) and remove the internal organs using a pair of forceps (Fig. 1d). Discard both head and internal organs and transfer the embryos to a new 90 mm bacteriological dish containing 10 mL of sterile PBS.
6. Mince the embryos into small pieces by passing them through a sterile 19 G needle three times in sterile PBS (*see* **Note 6**). Perform this step and all subsequent steps in an aseptic environment in a tissue culture hood to prevent contamination.
7. Transfer the minced embryo/PBS solution from a pool of three embryos (around 2 mL, *see* **Note 6**) to a 90 mm tissue culture dish containing 8 mL of sterile MEF medium.
8. Incubate the dish in a humidified incubator at 37 °C and 5 % CO_2 overnight.
9. Carefully replace the medium with 10 mL of fresh sterile MEF medium on the next day (*see* **Note 7**).

10. Incubate the dish in a humidified incubator at 37 °C and 5 % CO_2 overnight.
11. Passage the primary MEFs at a split ratio of 1:3 (*see* **Note 8**) on the next day: Wash the cells twice in PBS and add 3 mL of pre-warmed 0.05 % trypsin/EDTA solution (*see* **Note 9**) to a 90 mm tissue culture dish. Incubate the dish in a humidified incubator at 37 °C for 5 min. Tap the dish to confirm cell detachment and add 6 mL of sterile MEF medium. Transfer the cell suspension to a sterile 30 mL polypropylene tube and spin the cells down at 140×*g* for 3 min at room temperature. Aspirate the supernatant, resuspend the cell pellet in sterile MEF medium (10 mL for a 90 mm tissue culture dish), and seed MEFs onto a gelatin-coated 90 mm tissue culture dish (*see* **Note 10**). Incubate the dish in a humidified incubator at 37 °C and 5 % CO_2 overnight.
12. Expand MEFs to around passage 5 (*see* **Note 11**): When the culture reaches around 90 % confluence (every 3–4 days), passage the cells at a split ratio of 1:3 as in **step 11** above.
13. Treat a 90 mm tissue culture dish of MEFs (around 90 % confluence) at around passage 5 with 10 μg/mL mitomycin C: Combine 50 μL of sterile mitomycin C solution (1 mg/mL in PBS) with 5 mL of sterile MEF medium. Add this solution to a 90 mm dish of MEFs following aspiration of the culture medium. Incubate the dish in a humidified incubator at 37 °C and 5 % CO_2 for 2 h (*see* **Note 12**).
14. Freeze mitomycin C-treated MEFs ("feeder cells"): After 2 h of mitomycin C treatment, detach the cells as in **step 11** above. Gently resuspend the cell pellet in sterile freezing medium consisting of 10 % (v/v) DMSO, 10 % (v/v) FCS, and 80 % (v/v) MEF medium and quickly transfer the cell suspension into a labelled sterile cryovial. Freeze a 90 mm tissue culture dish of MEFs in 1 mL of freezing medium in one cryovial. Quickly place the cryovials into a cryo container and transfer to a −80 °C freezer overnight. Transfer the cryovials to a liquid nitrogen tank on the next day.

3.2 Derivation of EpiSCs from E5.5–E6.5 Postimplantation Epiblast

1. Thaw a vial of feeder cells into a 24-well dish 2 days prior to epiblast dissection: Place the cryovial into a 37 °C water bath until the ice dissolves. Quickly transfer the cell suspension into a 30 mL polypropylene tube containing 8 mL of sterile MEF medium and centrifuge at 140×*g* for 3 min at room temperature. Resuspend the cell pellet in 24 mL of sterile MEF medium and add 1 mL of cell suspension to each well of a 24-well dish (final feeder cell density: 3.5–4×10^4 cells per cm^2). Incubate the dish in a humidified incubator at 37 °C and 5 % CO_2 overnight. Replace the medium in each well with 1 mL of fresh sterile MEF medium on the next day. Incubate the dish in a humidified incubator at 37 °C and 5 % CO_2 until use.

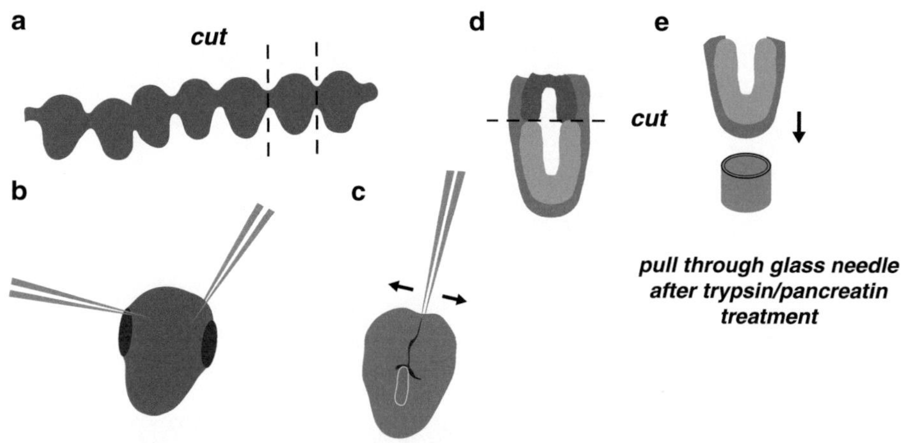

Fig. 2 Derivation of EpiSCs from E5.5–E6.5 postimplantation epiblast. The uterine wall between individual deciduae is cut (**a**), the embryo is dissected out of the decidua (**b**, **c**), and the trophectoderm is separated from the epiblast/visceral endoderm tissue (**d**). Next, the epiblast/visceral endoderm tissue is pulled through a fine glass needle following trypsin/pancreatin treatment to remove the visceral endoderm from the epiblast (**e**). Finally, the dissected epiblast is transferred to a dish containing sterile EpiSC medium and feeder cells and incubated in a humidified incubator at 37 °C and 5 % CO_2

2. Prepare fine glass needles of various diameters (90–250 μm) by pulling borosilicate glass capillaries and cutting the end (*see* **Note 13**).
3. Prepare sterile EpiSC medium on the day of epiblast dissection. Wash the wells of a 24-well dish containing feeder cells (*see* **step 1** above) once with PBS and add 500 μL of sterile EpiSC medium to each well.
4. Open the abdominal cavity and remove the uterus from two to three pregnant female mice at 5.5–6.5 days of gestation (preferably with *Oct4*-ΔPE-GFP reporter; *see* **Note 5**).
5. Place the uterus into a 90 mm bacteriological dish containing 10 mL of sterile MEF medium and cut the uterine wall between individual deciduae (Fig. 2a).
6. Open up the deciduae using a pair of forceps and carefully dissect out the embryos (Fig. 2b, c).
7. Transfer the embryos to a new 90 mm bacteriological dish containing 5 mL of sterile MEF medium.
8. Separate the trophectoderm from the epiblast/visceral endoderm tissue using a surgical blade (Figs. 2d and 3a, b).
9. Discard the trophectoderm and wash the epiblast/visceral endoderm tissue once in sterile PBS (*see* **Note 14**).
10. Place the epiblast/visceral endoderm tissue in sterile trypsin/pancreatin solution for 8 min at room temperature (*see* **Note 14**).
11. Transfer the epiblast/visceral endoderm tissue to a well of a 4-well dish containing 1 mL of EpiSC medium (*see* **Note 14**).

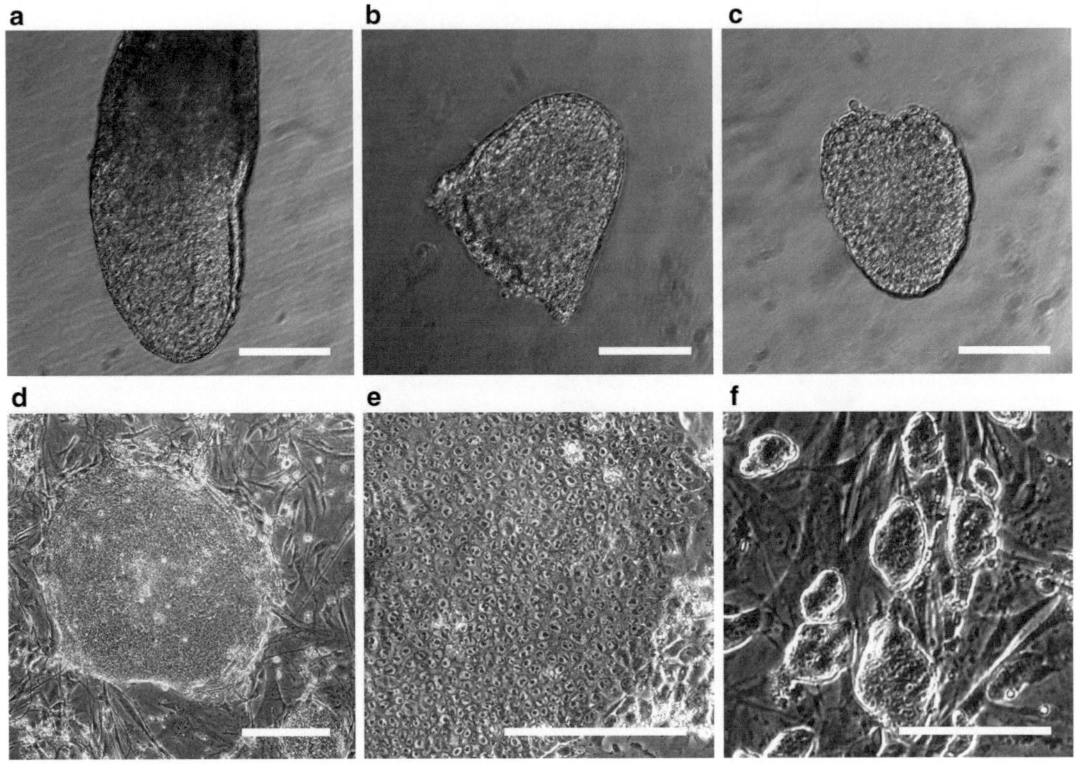

Fig. 3 Morphology of postimplantation epiblast, EpiSCs, and rESCs. Phase contrast images show an embryo at E6.5 (**a**), an embryo at E6.5 with the trophectoderm removed (**b**), and an embryo at E6.5 with both trophectoderm and visceral endoderm removed (**c**). Additional images show EpiSCs derived from E6.5 epiblast and maintained in EpiSC medium (activin A, bFGF, knockout serum replacement) on feeder cells (**d**, **e**), and rESCs grown in MEF medium (FCS, LIF) on feeder cells (**f**). Scale bar: 100 μm

12. Pull the epiblast/visceral endoderm tissue through a fine glass needle to remove the visceral endoderm from the epiblast (*see* **Note 15**, Fig. 2c, e).
13. Discard the visceral endoderm and transfer the dissected epiblast into a well of a 24-well dish containing 500 μL of sterile EpiSC medium and feeder cells (*see* **steps 1** and **3** above).
14. Incubate the dish in a humidified incubator at 37 °C and 5 % CO_2 overnight.
15. Replace the medium with 500 μL of fresh sterile EpiSC medium every day. In addition, monitor the colony size and morphology every day (*see* **Note 16**). Perform this step and all subsequent steps in an aseptic environment in a tissue culture hood to prevent contamination.
16. Passage the primary colony after 5–10 days (*see* **Note 17**) at a split ratio of 1:1. Gently aspirate the culture medium and add 500 μL of collagenase solution (pre-warmed to room temperature) to a well of a 24-well dish. Incubate the colony in collagenase for 6–8 min at room temperature. In the meantime

wash a well of a 24-well dish containing feeder cells (*see* **step 1** above) once with PBS and change the medium to 500 μL of fresh sterile EpiSC medium. After 6–8 min of incubation in collagenase, carefully wash the colony twice in sterile DMEM/F-12 basal medium and add 500 μL of sterile EpiSC medium to the well. Pipet up and down using a 200 μl pipette under a stereomicroscope to break the colony into small clumps (*see* **Note 18**). Transfer the clumps into a well of a 24-well dish containing 500 μL of EpiSC medium and feeder cells. Incubate the dish in a humidified incubator at 37 °C and 5 % CO_2 overnight.

17. Replace the medium with 500 μL of fresh sterile EpiSC medium every day (*see* **Note 19**). In addition, monitor the colony size and morphology every day (*see* **Note 20**, Fig. 3d, e).
18. Passage the cells after 3–6 days at a split ratio of 1:2–1:4 as in **step 16** above.
19. Expand the cells to a 6-well dish (*see* **Note 21**) and freeze several vials of EpiSCs in sterile freezing medium consisting of 10 % (v/v) DMSO and 90 % (v/v) knockout serum replacement: Detach the cells as in **step 16** above (use 1.5 mL of collagenase per well of a 6-well dish), transfer the dissociated EpiSC colonies into a 30 mL polypropylene tube containing 8 mL of sterile DMEM/F-12 basal medium, and spin the cells down at 140 × *g* for 3 min at room temperature. Gently resuspend the cell pellet in freezing medium and transfer the cell suspension into a labelled sterile cryovial. Freeze one well of EpiSCs in a 6-well dish in 0.5 mL of freezing medium in one cryovial. Place the cryovials into a cryo container and transfer to a −80 °C freezer overnight. Transfer the cryovials to a liquid nitrogen tank on the next day. The content of one cryovial should be thawed into one well of a 6-well dish (*see* **Note 22**).

3.3 Reversion of EpiSCs to rESCs Upon Exposure to LIF/Stat3 Signalling on Feeder Cells

1. Thaw a vial of feeder cells into several wells of a 12-well dish 5–7 days prior to the reversion experiment (*see* **Note 23**): Place the vial into a 37 °C water bath until the ice dissolves. Quickly transfer the cell suspension into a 30 mL polypropylene tube containing 8 mL of sterile MEF medium and centrifuge at 140 × *g* for 3 min at room temperature. Resuspend the cell pellet in sterile MEF medium and add 2 mL of the cell suspension per well of a 12-well dish. Incubate the dish in a humidified incubator at 37 °C and 5 % CO_2 overnight. Replace the medium in each well with 2 mL of fresh sterile MEF medium on the next day. Incubate the dish in a humidified incubator at 37 °C and 5 % CO_2 until use.
2. Prepare sterile ESC medium on the day of the reversion experiment. Wash a well of a 12-well dish containing feeder

cells (*see* **step 1** above) once with PBS and change the medium to 2 mL of sterile ESC medium.

3. Detach a sub-confluent well of EpiSCs (preferably with *Oct4-ΔPE-GFP* reporter, *see* **Note 5**) in a 12-well dish using 1 mL of collagenase as in **step 16** in Subheading 3.2.
4. Transfer the dissociated EpiSC colonies into a 30 mL polypropylene tube containing 8 mL of sterile DMEM/F-12 basal medium and spin the cells down at 140 × *g* for 3 min at room temperature. Aspirate the supernatant and resuspend the colonies in 800–1,000 μL of sterile ESC medium. Transfer 100 μL of cell suspension into a well of a 12-well dish (*see* **Note 24**) containing 2 mL of sterile ESC medium and 5–7-day-old feeder cells (*see* **steps 1** and **2** above).
5. Incubate the dish in a humidified incubator at 37 °C and 5 % CO_2 overnight.
6. Replace the medium with 2 mL of fresh sterile ESC medium every day for about 15–25 days.
7. After 5 days of reversion, check colonies for GFP expression in a fluorescent microscope every day. GFP-positive colony patches typically appear on days 10–20 (*see* **Note 25**). Depending on the growth rate, it may be necessary to passage the cells once prior to appearance of GFP-positive cells (on days 7–9), using collagenase as in **step 16** in Subheading 3.2 and at a split ratio of around 1:2.
8. Pick colony patches with more than 200 GFP-positive cells after 15–25 days of reversion using a fine glass needle (*see* **Note 26**) under the fluorescent microscope.
9. Transfer the GFP-positive colony patches to a new well of a 12-well dish containing 2 mL of sterile ESC medium and 5–7-day-old feeder cells (*see* **steps 1** and **2** above).
10. Incubate the dish in a humidified incubator at 37 °C and 5 % CO_2 overnight.
11. Replace the medium with 2 mL of fresh sterile ESC medium every day.
12. Thaw a vial of feeder cells into several wells of a 6-well dish (3 mL of MEF medium per well) as in **step 1** above.
13. Dissociate GFP-positive colonies (100–200 μm diameter) to a single-cell suspension using 0.05 % trypsin/EDTA 3–5 days after colony picking (*see* **steps 8** and **9** above): Wash the cells twice in PBS and add 500 μL of pre-warmed trypsin/EDTA solution (*see* **Note 9**) to a well of a 12-well dish. Incubate the dish in a humidified incubator at 37 °C for 8–10 min. Tap the dish to confirm cell detachment and add 1 mL of sterile MEF medium to the dish to inactivate the trypsin. Transfer the cell suspension into a sterile 30 mL polypropylene tube containing

5 mL of sterile MEF medium and spin the cells down at 140 × *g* for 3 min at room temperature. Aspirate the supernatant and resuspend the cell pellet in 1 mL of sterile ESC medium. Transfer the cell suspension to a well of a 6-well dish containing feeder cells (*see* **step 12** above) and 3 mL of sterile ESC medium. Incubate the dish in a humidified incubator at 37 °C and 5 % CO_2 overnight. These cells can now be expanded as rESCs (*see* **Note 27**, Fig. 3f).

14. Freeze rESCs in sterile freezing medium containing 10 % (v/v) DMSO, 10 % (v/v) FCS, and 80 % (v/v) ESC medium: Detach the cells as in **step 13** above, gently resuspend the cell pellet in freezing medium, and transfer the cell suspension into a labelled cryovial. Freeze one well of rESCs in a 6-well dish in 1 mL of freezing medium in one cryovial. Place the cryovials into a cryo container and transfer to a −80 °C freezer overnight. Transfer the cryovials to a liquid nitrogen tank on the next day.

4 Notes

1. Dissolve 80 mg of adenosine, 85 mg of guanosine , 73 mg of cytidine, 73 mg of uridine, and 24 mg of thymidine (all from Sigma) in 100 mL of dH_2O by heating on a hot plate to about 40 °C. Filter through a 0.45 μm membrane, aliquot, and store at −20 °C for up to 1 year.
2. Test several batches of FCS for ESC derivation and culture, since batches vary in quality. Select the batch with the highest ESC derivation efficiency for use in EpiSC reversion experiments.
3. Calcium/magnesium-free Tyrode Ringer's Saline pH 7.6–7.7 is composed of 136.9 mM NaCl, 4 mM KCl, 0.15 mM $NaH_2PO_4{\cdot}5H_2O$, 0.2 mM KH_2PO_4, 11.9 mM $NaHCO_3$, and 11 mM glucose. Dissolve 8 g NaCl, 0.3 g KCl, 0.093 g $NaH_2PO_4{\cdot}5H_2O$, 0.025 g KH_2PO_4, 1 g $NaHCO_3$, and 2 g glucose in 1,000 mL of dH_2O. Adjust pH to 7.6–7.7.
4. A mouth pipette consists of an aspirator mouthpiece, tubing, and a Pasteur pipette [9]: Pull a Pasteur pipette on a flame to create a narrow opening. Attach the Pasteur pipette to a 1,000 μL tip with an inserted cotton plug and connect it to a mouthpiece through tubing. Insert a sterile 0.45 μm filter in the middle part of the tubing. As an alternative to a Pasteur pipette, a fine glass needle (*see* **Note 13**) can be mounted onto the mouth pipette.
5. The *Oct4*-ΔPE-GFP reporter enables rapid identification of rESCs, since GFP expression reflects activation of the distal *Oct4* enhancer that is inactive in EpiSCs and gets activated upon reversion to rESCs [7, 8]. Note that the genetic background of the mice used for EpiSC derivation has an impact on

derivation efficiency, quality of EpiSC lines, and their efficiency of reversion to rESCs. The use of a permissive background such as 129 enables robust reversion of EpiSCs to rESCs [10].

6. Attach a sterile 19 G needle to a sterile 5 mL syringe. Remove the plunger and place three embryos in the syringe. Add 2 ml of sterile PBS and insert the plunger. Pass the embryo/PBS solution through the needle three times.
7. Carefully remove the supernatant and add MEF medium as gently as possible to avoid detachment of tissue pieces and thus low yield of primary MEFs.
8. Transfer the cells of one confluent 90 mm tissue culture dish to three new 90 mm dishes.
9. Pre-warm the trypsin/EDTA solution to 37 °C in a water bath, since incubation in cold solution results in slower cell detachment, longer incubation times, and potentially more cell death.
10. Coat a 90 mm tissue culture dish with 5 mL of sterile 0.1 % gelatin solution for 30–60 min at room temperature. Aspirate the gelatin solution and dry the dish before use.
11. Expand primary MEFs to a maximum of five passages for their use as feeder cells, because their quality decreases with increasing passage number. Old cells may produce different cytokines and display altered adhesive properties compared to young cells. Alternatively, primary MEFs can be frozen at an early passage (*see* **step 14** in Subheading 3.1) and stored in liquid nitrogen. The cells can be thawed anytime (*see* **step 1** in Subheading 3.2), expanded, and treated with mitomycin C (*see* **step 13** in Subheading 3.1). The mitomycin C-treated MEFs ("feeder cells") can be used immediately or frozen in liquid nitrogen for later use.
12. Do not incubate MEFs in mitomycin C solution for more or less than 2 h. Longer and shorter incubation times result in high toxicity and incomplete inhibition of cell proliferation, respectively.
13. Pull borosilicate glass capillaries using a micropipette puller (settings for P-80 Brown-Flaming Micropipette Puller/Sutter Instruments: HEAT = 550, PULL = 50, VEL = 50). Prepare glass needles of various diameters (90–250 μm) by cutting the end of the needles.
14. Use a mouth pipette (*see* **Note 4**) with a mounted glass needle (*see* **Note 13**; 250 μm diameter) to transfer the epiblast/visceral endoderm tissue to a well of a 4-well dish containing 1 mL of sterile PBS, trypsin/pancreatin solution, or EpiSC medium. Epiblasts tend to be sticky following removal of the extraembryonic tissues and they occasionally attach to the surface of the glass needle. To avoid loss of epiblasts during

the transfer, suck in enough liquid and transfer the epiblasts as quickly as possible.

15. Use a mouth pipette (*see* **Note 4**) with a mounted glass needle (*see* **Note 13**) to suck the epiblast/visceral endoderm tissue from the distal end into the narrow opening of the glass needle (Fig. 2e). Depending on the size of the epiblast/visceral endoderm tissue, use glass needles with a diameter of 90–250 μm. Start with a wide glass needle, e.g., 250 μm. If the visceral endoderm does not separate from the epiblast, repeat this step using a narrow glass needle, e.g., 90 μm. The use of a glass needle with a diameter that is less than that of the epiblast (Fig. 2e) facilitates removal of the visceral endoderm.
16. The dissected epiblast attaches to the feeder-coated surface of the dish overnight and spreads out quickly to form a primary colony. The primary colony is compact and flat, and the cells show a high nuclear-to-cytoplasmic ratio (Fig. 3e). Over the next few days the colony grows and cells at the edge and the center start to differentiate. Cell death is common at this stage.
17. Passage the primary colony when the cells in the center of the colony die or differentiate. Do not wait until the entire colony differentiates (i.e., colony gets less compact with individual cells flattening out).
18. An optimal clump size (200–400 cells per clump) is crucial for the success of the cultures. Unlike ESCs, EpiSCs are sensitive to dissociation into single cells. If the clumps are too small or too big, the cells will die or differentiate, respectively. As an alternative to pipetting, the primary colony can be cut into small pieces using a fine glass needle (*see* **Note 26**).
19. The high proliferation rate of EpiSCs necessitates a frequent change of culture medium. As an alternative to a daily medium change, the pH value of the medium can be used as an indicator: a yellow color of EpiSC medium (i.e., DMEM/F-12 with phenol red) indicates a drop in pH value, requiring a medium change. In addition, EpiSCs must be incubated in a saturated humidified atmosphere to mimimize media evaporation.
20. EpiSCs form flattened, yet compact, colonies (Fig. 3d) and show a high nuclear-to-cytoplasmic ratio and prominent nucleoli (Fig. 3e) [4, 5]. Differentiated cells are usually present at the edge of the colony (Fig. 3d, e), especially at early passage numbers. The colony compactness and morphology of EpiSC lines vary and may be influenced by genetic background. The derivation efficiency of EpiSCs from E6.5 epiblast is generally around 70 % (i.e., 7 out of 10 explanted epiblasts give rise to a continuously growing EpiSC line).
21. Passage EpiSCs every 1–2 days at a split ratio of 1:2–1:4, since frequent passage and high plating densities minimize cell death

and differentiation. The growth rate of different EpiSC lines varies slightly and may depend on genetic background. EpiSCs can be maintained on feeder cells or moved to feeder-free conditions on fibronectin. As an alternative to dissociation of EpiSC colonies with collagenase and pipetting, EpiSCs can be detached using accutase (incubation for 1–2 min at 37 °C). Since accutase treatment generates a near-single-cell suspension, but is less toxic than trypsin/EDTA, this method is particularly suitable for flow cytometric sorting and for feeder-free maintenance of EpiSCs. Once established, verify pluripotency gene expression of EpiSC lines by immunostaining with Oct4, Sox2, and Nanog antibodies [4]. Confirm the presence of nuclear foci of histone 3 lysine 27 trimethylation (H3K27me3), indicating X-chromosome inactivation, in female EpiSC lines by immunostaining with H3K27me3 antibodies [6]. Determine the karyotype of EpiSCs, e.g., by metaphase spreads, to detect chromosomal abnormalities [4]. Test the differentiation potential of EpiSCs by teratoma formation or directed in vitro differentiation assays [4, 5].

22. Place a cryovial of EpiSCs into a 37 °C water bath until the ice dissolves. Quickly transfer the cell suspension into a 30 mL polypropylene tube containing 8 mL of sterile DMEM:F-12 basal medium and centrifuge at 140 × *g* for 3 min at room temperature. Gently resuspend the cell pellet in 3 mL of sterile EpiSC medium and transfer the cell suspension to a well of a 6-well dish containing feeder cells after one wash in sterile PBS. Incubate the dish in a humidified incubator at 37 °C and 5 % CO_2 overnight. To maximize plating efficiency, EpiSCs should be plated at high density after thawing and passaged on the day after thawing.

23. An optimal feeder cell density ($7–7.5 \times 10^4$ cells per cm^2) is crucial for the success of the reversion experiment. The feeder cell density should be higher than the density used for EpiSC derivation.

24. Plate EpiSC colonies at low density, since the reversion process is slow (15–25 days). A subconfluent well of EpiSCs in a 12-well dish should be seeded in 8–10 wells of a 12-well dish (i.e., split ratio of 1:8–1:10).

25. The success, rate, and efficiency of reversion of EpiSCs to rESCs depend on several factors, such as the quality of the FCS batch (*see* **Note 2**), the quality of feeder cells (*see* **Note 11**), the density of feeder cells (*see* **Note 23**), the quality of EpiSCs (*see* **Notes 20** and **21**), the density of EpiSCs (*see* **Note 24**), and the genetic background (*see* **Note 5**). In general, reversion of EpiSCs to rESCs takes between 2 and 3 weeks and several GFP-positive colony patches appear per well of a 12-well dish.

26. Pull borosilicate glass capillaries using a micropipette puller (settings for P-80 Brown-Flaming Micropipette Puller/Sutter Instruments: HEAT = 550, PULL = 50, VEL = 50), but do not cut the end of the glass needles for this application.
27. The growth rate and morphology of rESCs are similar to mouse ESCs: Passage rESCs every 2–3 days at a split ratio of 1:10–1:20. The cells show a dome-shaped morphology (Fig. 2.3f) with homogeneous *Oct4*-ΔPE-GFP expression [7] when grown in MEF medium (FCS and LIF) on feeder cells. Verify pluripotency gene expression of rESCs by immunostaining with Oct4, Sox2, and Nanog antibodies and confirm the loss of nuclear foci of H3K27me3, that indicates X-chromosome reactivation, in female rESCs by immunostaining with H3K27me3 antibodies [7]. Inject 15–20 *Oct4*-ΔPE-GFP rESCs into E3.5 C57BL/6 blastocysts and assess chimerism by agouti coat color of pups [7]. Confirm germ line contribution by inspecting E13.5 C57BL/6 male and female genital ridges for GFP-positive cells and by assessing the coat color of progeny derived from chimeras [7].

Acknowledgement

This work was supported by grants from the Wellcome Trust to A.G (RG44593) and to M.A.S.

References

1. Nichols J, Smith A (2009) Naive and primed pluripotent states. Cell Stem Cell 4:487–492
2. Borgel J, Guibert S, Li Y et al (2010) Targets and dynamics of promoter DNA methylation during early mouse development. Nat Genet 42:1093–1100
3. Monk M, Harper MI (1979) Sequential X chromosome inactivation coupled with cellular differentiation in early mouse embryos. Nature 281:311–313
4. Brons IG, Smithers LE, Trotter MW et al (2007) Derivation of pluripotent epiblast stem cells from mammalian embryos. Nature 448:191–195
5. Tesar PJ, Chenoweth JG, Brook FA et al (2007) New cell lines from mouse epiblast share defining features with human embryonic stem cells. Nature 448:196–199
6. Guo G, Yang J, Nichols J et al (2009) Klf4 reverts developmentally programmed restriction of ground state pluripotency. Development 136:1063–1069
7. Bao S, Tang F, Li X et al (2009) Epigenetic reversion of post-implantation epiblast to pluripotent embryonic stem cells. Nature 461: 1292–1295
8. Yeom YI, Fuhrmann G, Ovitt CE et al (1996) Germline regulatory element of Oct-4 specific for the totipotent cycle of embryonal cells. Development 122:881–894
9. Nagy A (2003) Manipulating the mouse embryo: a laboratory manual, 3rd edn. Cold Spring Harbor Laboratory, Cold Spring Harbor, NY
10. Hanna J, Markoulaki S, Mitalipova M et al (2009) Metastable pluripotent states in NOD-mouse-derived ESCs. Cell Stem Cell 4:513–524

Chapter 3

Converting Mouse Epiblast Stem Cells into Mouse Embryonic Stem Cells by Using Small Molecules

Wenlin Li and Sheng Ding

Abstract

The derivation of mouse embryonic stem cells (mESCs) from the inner cell mass (ICM) of mouse preimplantation blastocysts has provided tremendous opportunities in exploring mammalian development and pluripotent stem cell biology (Martin, Proc Natl Acad Sci U S A 78(12):7634–7638, 1981; Evans and Kaufman, Nature 292(5819):154–156, 1981). The progresses of mESC studies also enabled the subsequent establishment of human ESCs (hESCs) from human embryos (Thomson et al., Science 282(5391):1145–1147, 1998). However, it had been a long-standing mystery why mESCs and hESCs exhibit distinct differences in terms of self-renewal signaling dependence and cell culture behaviors, such as colony morphology and response to single-cell dissociation, although both of them are pluripotent. Recently, a novel type of pluripotent stem cells were isolated from the postimplantation egg cylinder-stage epiblasts of mouse and rat (Tesar et al., Nature 448(7150):196–199, 2007; Brons et al., Nature 448(7150):191–195, 2007), and were named epiblast stem cells (EpiSCs). Interestingly, mouse EpiSCs (mEpiSCs) closely resemble hESCs in respect of the colony morphology and self-renewal dependence. These results suggest that different pluripotent states exist during mammalian development: the mESC-like state representing the preimplantation ICM and hESC/mEpiSC-like state representing later epiblast cells. This chapter introduces methods of converting mEpiSCs back to mESCs through chemical approaches.

Key words Embryonic stem cells, Epiblast stem cells, Small molecules

1 Introduction

Mouse embryonic stem cells (mESCs) are pluripotent cells derived from inner cell mass (ICM) of preimplantation blastocysts [1, 2]. Recently, a distinct type of pluripotent stem cells were isolated from the postimplantation egg cylinder-stage epiblasts of mouse and rat [3, 4]. They were named epiblast stem cells (EpiSCs). Interestingly, mouse EpiSCs (mEpiSCs) closely resemble human ESCs (hESCs) in respect of the colony morphology and self-renewal dependence [5]. Unlike mESCs growing in three-dimensional, domed colonies in culture, mEpiSCs form flattened two-dimensional colonies and survive poorly after dissociation into single cells. In addition, the essential self-renewing signaling pathways for mESCs, such as leukemia

Ramiro Alberio (ed.), *Epiblast Stem Cells: Methods and Protocols*, Methods in Molecular Biology, vol. 1074, DOI 10.1007/978-1-62703-628-3_3,

inhibitory factor (LIF)/signal transducers and activators of transcription (STAT3) and bone morphogenetic protein (BMP)/Smad [6–8], do not support mEpiSC self-renewal. However, similar to hESCs, bFGF and activin A/nodal signaling are essential to support the undifferentiated state of mEpiSCs. mEpiSCs can generate cell types in all three germ layers and form teratoma after transplantation. However, their ability to contribute to chimeras is poor after grafted into mouse preimplantation blastocysts. Latest studies demonstrated that mEpiSCs readily generated chimeras with contribution to all three embryonic germ layers and primordial germ cells when transplanted to postimplantation embryos (embryonic day 7.5) in whole-embryo culture. However, transplanted mESCs have little chimerism competence in embryonic day 7.5 embryos [9]. These discrepancies between mEpiSCs and mESCs enforced the notion that they represent different pluripotent stages of development: the mESC-like state representing the preimplantation ICM and hESC/mEpiSC-like state representing later epiblast cells. By combining genetic reprogramming and cell signaling modulation by small molecules that favor the mESC-like state, mESC-like human induced pluripotent stem cells (m-hiPSCs) can be generated from human fibroblasts by ectopic expression of Oct4, Sox2, Nanog, and Lin28 in culture medium that contains human LIF and the combination of three chemical inhibitors of activin receptor-like kinase 5 (ALK5), MEK, and glycogen synthase kinase 3 (GSK-3) [10]. m-hiPSCs form domed colonies and exhibit stable long-term self-renewal when cultured with hLIF and the small molecule combination, suggesting that m-hiPSCs are distinct from hESCs in signaling responses and self-renew in the mESC-like state. The protocol to induce m-hiPSCs was previously published in another MiMB book [11].

However, currently, authentic ESCs representing ICM-stage pluripotent cells have only been derived from rodents. Although ESCs from primates, including human, were also derived from starting with preimplantation blastocysts, they appear to correspond to the late-epiblast-stage pluripotent cells. Better understanding of different pluripotent stages would be essential to derive ICM-stage ESCs from other species besides rodents. In our previous study, we found that blockage of the TGFβ pathway or inhibition of histone demethylase LSD1 (lysine-specific histone demethylase) with small-molecule inhibitors induced dramatic morphological changes in EpiSCs toward mESC phenotypes with simultaneous activation of ICM-specific gene expression. However, full conversion of EpiSCs to the mESC-like state with chimerism competence could be readily generated only with the combination of inhibitors of LSD1, ALK5, MEK, fibroblast growth factor receptor (FGFR), and GSK-3. In this chapter, we introduce methods of converting mEpiSCs to mESCs through synergistically modulating related signaling pathways and epigenetic mechanisms by chemical approaches without any genetic manipulations.

2 Materials

2.1 Cells and Cell Culture Reagents

1. The mEpiSC line (line EpiSC-5, male): Gift from Dr. Paul Tesar (Case Western Reserve University).
2. mEpiSC medium: Dulbecco's modified Eagle's medium/F12 (DMEM/F12) with 10 mM Hepes, 20 % knockout serum replacement (KSR), 1 % glutamax, 1 % nonessential amino acids, 1 % penicillin/streptomycin, 0.1 mM β-mercaptoethanol, and 10 ng/ml bFGF.
3. mESC medium on irradiated mouse embryonic fibroblasts (MEFs): Knockout DMEM supplemented with 20 % KSR, 0.1 mM β-mercaptoethanol, 2 mM L-glutamine, 0.1 mM non-essential amino acids, and 10^3 units/ml recombinant murine leukemia inhibitory factor (LIF) (ESGRO, Millipore, CA, USA).
4. Cell dissociation reagents: 0.05 % Trypsin–EDTA and 1 mg/ml collagenase type IV (dissolved in DMEM/F12).
5. The medium for MEFs: DMEM (high glucose) supplemented with 10 % FBS, 2 mM L-glutamine, and 0.1 mM 2-mercaptoethanol.
6. Small molecular inhibitors: GSK3 inhibitor (CHIR99021, Stemgent, CA, USA), MEK inhibitor (PD0325901, Stemgent), ALK5 inhibitors (A-83-01 and SB431542, Stemgent), FGF receptor inhibitor (PD173074, Stemgent), and LSD1 inhibitor (Parnate, Sigma, MO, USA) (*see* **Note 1**). Stored in aliquots at −20 °C.
7. Matrigel (BD Biosciences, CA, USA).
8. All cell culture products were purchased from Invitrogen (CA, USA) except where mentioned.

2.2 Cytochemistry and Immuno-fluorescence Assay

1. Alkaline phosphatase staining was performed using the alkaline phosphatase kit (Sigma-Aldrich).
2. Fixation buffer for immunofluorescence: 4 % paraformaldehyde (Sigma-Aldrich) in PBS (Invitrogen).
3. Washing buffer: PBS containing 0.1 % Triton X-100 (Sigma-Aldrich).
4. Blocking buffer: 0.1 % Triton X-100 and 10 % normal donkey serum (Jackson ImmunoResearch Laboratories Inc, UK) in PBS.
5. The primary antibodies (vendors and dilution in the bracket): Albumin (Abcam, AB19188, 1:200); Brachyury (Santa Cruz, C-19, 1:200); Cardiac troponin T (CT3) (Developmental Studies Hybridoma Bank, 1:700); MAP2 (Abcam, ab5392, 1:1,000); MF20 (Developmental Studies Hybridoma Bank, 1:200); Nanog (Abcam, ab21603, 1:500); Oct4 (Santa Cruz,

sc-5279, 1:100); Sox17 (R&D systems, AF1924, 1:300); SSEA1 (Santa Cruz, sc-21702, 1:100); Stella (Millipore, MAB4388, 1:200); Tuj-1 (Covance, MMS-435P, 1:1,000).

6. Secondary antibodies: Alexa Fluor 486/555 donkey anti-mouse, anti-goat, or anti-rabbit IgG (1:1,000) (Invitrogen).
7. DAPI (Sigma-Aldrich) is dissolved in tissue-culture water at 1 mg/ml, stored in aliquots at −20 °C, and used at 10 μg/ml.
8. Nikon Eclipse TE2000-U microscope.

2.3 Semiquantitative RT-PCR and Real-Time PCR

1. RNeasy Plus Mini Kit and QIAshredder (Qiagene, CA, USA).
2. iScript™ cDNA Synthesis Kit and iQ SYBR Green Supermix (Bio-Rad, CA, USA).
3. Platinum PCR SuperMix (Invitrogen).
4. Primers used were as reported [12].

2.4 Bisulfite Sequencing Analysis

1. Non-Organic DNA Isolation Kit (Millipore).
2. EZ DNA Methylation-Gold Kit (Zymo Research Corp. CA, USA).
3. Primers used for Fgf4 and Stella promoter fragment were as reported [13, 14]: Primers for Fgf4 (TTT AGG TTT TAA GAG TGT TGG GGA GAA GAT and TAC AAA ACA AAA ACA TCA AAC CCA TTC TAA), and nested PCR primers for Stella promoter fragment (outer primer: ATT TTG TGA TTA GGG TTG GTT TAG AA, and CCA AAA CAT CCT CTT CAT CTT TCT TCT and inner primer: TTT TTG GAA TTG GTT GGG ATT G, and CTT CTA AAA AAT TTC AAA ATC CTT CAT T).

3 Methods

3.1 Culture and Passage of mEpiSCs

1. MEFs were irradiated with 30–60 gray and were seeded as feeders at a density of 5×10^4 cells/cm^2.
2. EpiSCs were cultured on feeders with mEpiSC medium. The medium was changed every day. The cells were passaged by 1:6 when reaching about 60–70 % confluence.
3. EpiSCs were treated with 1 mg/ml collagenase type IV for about 5–15 min at 37 °C until the edges of colonies begin to pull away from the plate (*see* **Note 2**). After aspirating the collagenase type IV solution, the cells were gently detached from plate by washing with mEpiSC medium. Transfer the detached cells into a 15 ml tube, and spin them down by centrifuging at $300 \times g$ for 5 min. Resuspend the cells by appropriate volume of mEpiSC medium, gently pipette up and down 1–2 times to break up cells into small clumps of 10–100 cells, and then plate cells onto new plates/dishes with feeders.

3.2 Convert mEpiSCs to the mESC-Like Pluripotent State by Small Molecules

1. EpiSCs were trypsinized to single cells, and about 500 cells per well were plated into 6-well plate and cultured with mESC medium supplemented with 2 μM Parnate, a small molecular LSD1 inhibitor, for 4 days. Under such condition, about 70–80 % of the EpiSCs formed small and compact colonies similarly to mESCs.
2. After 4-day Parnate treatment, the culture was selected with mESC medium supplemented with 0.5 μM MEK inhibitor PD0325901 and 0.1 μM FGFR inhibitor PD173074 for another 4 days. About 20 % of the cells survived the selection as domed and ALP-positive colonies.
3. The whole cell population was then passaged and expanded with inhibitors of MEK (PD0325901), FGFR (PD173074), ALK5 (A-83-01), and GSK3 (CHIR99021).
4. Individual colonies were picked up and expanded by the above condition for another 2-week culture. Stable cultures could be established and were morphologically indistinguishable from mESCs (Fig. 1).

3.3 Characterize the Converted mEpiSCs

1. The cells were fixed by 4 % paraformaldehyde in PBS, and washed three times by PBS containing 0.1 % Triton X-100. After blocking the cells by PBS with 0.1 % Triton X-100 and 10 % normal donkey serum, the cells were incubated with primary antibodies by indicated dilution overnight at 4 °C. After washing by PBS with 0.1 % Triton X-100 three times (5 min each time), the antigens were revealed by appropriate fluorescence-conjugated secondary antibodies. Then the cells were washed three times, 5 min per time, by PBS with 0.1 % Triton X-100. Immunocytochemistry should confirm homogeneous expression of pluripotency-associated markers in long-term expanded converted mEpiSCs, including Oct4, Nanog, SSEA1, and STELLA. STELLA is a protein which is expressed in mESCs but absent in mEpiSCs.
2. RNA from mEpiSCs and their converted cells was extracted by using the RNeasy Plus Mini Kit in combination with QIAshredder. Reverse transcription was performed with 1 μg RNA using iScript™ cDNA Synthesis Kit according to the manufacturer's instructions. The gene expression was analyzed by RT-PCR using Platinum PCR SuperMix. PCR products were resolved on 1.5 % agarose gels and visualized by ethidium bromide staining. RT-PCR analysis demonstrated restoration of gene expression of specific ICM and germline competence markers by converted cells, including *Dax1*, *Esrrb*, *Fbxo15*, *Fgf4*, *Pecam1*, *Rex1*, *Stella*, and *Stra8*. However, the genes associated with the EpiSCs and early germ layers, such as Fgf5 and Brachyury, were decreased or undetectable.

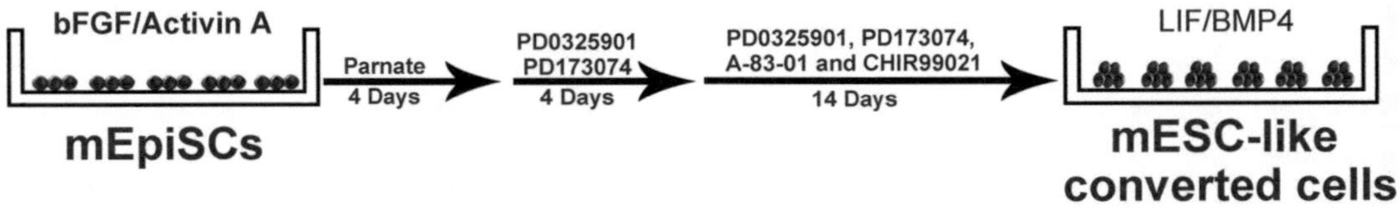

Fig. 1 Schematic for the generation of mouse embryonic stem cell (mESC)-like converted cells from mouse epiblast stem cells (mEpiSCs). Parnate, LSD1 inhibitor; PD0325901, MEK inhibitor; PD173074, FGFR inhibitor; A-83-01, ALK5 inhibitor; CHIR99021, GSK3 inhibitor

3. DNAs from mESCs, mEpiSCs, and converted cells were isolated using the Non-Organic DNA Isolation Kit. The DNAs were then treated for bisulfite sequencing with the EZ DNA Methylation-Gold Kit. The treated DNAs were used to amplify sequences of *Fgf4* and *Stella* promoter. The resulting fragments were cloned using the TOPO TA Cloning Kit and sequenced. The promoter regions of *Stella* and *Fgf4* gene in converted cells were largely unmethylated as mESCs, but were hypermethylated in EpiSCs.
4. For in vitro differentiation of converted cells, cells were trypsinized into single cells and cultured in suspension to form embryoid bodies (EBs) in ultralow attach plates in DMEM medium supplemented with 10 % FBS. After 6-day culture in suspension, EBs were replated onto 0.1 % gelatin-coated plates. Spontaneous differentiation was examined by immunostaining at various time points (3 up to 16 days). The cells were fixed for immunocytochemistry analysis as described. Mesoderm marker Brachyury should be detected at 3 days after plating EBs. Endoderm and ectoderm markers, such as albumin and βIII-tubulin, should be detected 7 days later.
5. The GFP-labeled converted cells were aggregated with 8-cell-stage mouse embryos, and were then transplanted into the uteri of 2.5 dpc pseudopregnant CD1 mice. Chimeras could be obtained with the presence of GFP integration in multiple adult tissues, including gonad.

4 Notes

1. The media containing small molecules cannot be filtered. Always add small-molecule stock solution to already sterilized media.
2. The cells should be tightly packed within the colony, and maintain a defined colony border. Partial differentiated cells with undefined border were difficult to be detached by collagenase type IV.

References

1. Martin GR (1981) Isolation of a pluripotent cell line from early mouse embryos cultured in medium conditioned by teratocarcinoma stem cells. Proc Natl Acad Sci U S A 78(12): 7634–7638
2. Evans MJ, Kaufman MH (1981) Establishment in culture of pluripotential cells from mouse embryos. Nature 292(5819):154–156
3. Tesar PJ, Chenoweth JG, Brook FA, Davies TJ, Evans EP, Mack DL, Gardner RL, McKay RD (2007) New cell lines from mouse epiblast share defining features with human embryonic stem cells. Nature 448(7150):196–199
4. Brons IG, Smithers LE, Trotter MW, Rugg-Gunn P, Sun B, de Sousa C, Lopes SM, Howlett SK, Clarkson A, Ahrlund-Richter L, Pedersen RA, Vallier L (2007) Derivation of pluripotent epiblast stem cells from mammalian embryos. Nature 448(7150):191–195
5. Thomson JA, Itskovitz-Eldor J, Shapiro SS, Waknitz MA, Swiergiel JJ, Marshall VS, Jones JM (1998) Embryonic stem cell lines derived from human blastocysts. Science 282(5391): 1145–1147
6. Niwa H, Burdon T, Chambers I, Smith A (1998) Self-renewal of pluripotent embryonic stem cells is mediated via activation of STAT3. Genes Dev 12(13):2048–2060
7. Ying QL, Nichols J, Chambers I, Smith A (2003) BMP induction of Id proteins suppresses differentiation and sustains embryonic stem cell self-renewal in collaboration with STAT3. Cell 115(3):281–292
8. Qi X, Li TG, Hao J, Hu J, Wang J, Simmons H, Miura S, Mishina Y, Zhao GQ (2004) BMP4 supports self-renewal of embryonic stem cells by inhibiting mitogen-activated protein kinase pathways. Proc Natl Acad Sci U S A 101(16):6027–6032
9. Huang Y, Osorno R, Tsakiridis A, Wilson V (2012) In vivo differentiation potential of epiblast stem cells revealed by chimeric embryo formation. Cell Rep 2(6):1571–1578
10. Li W, Wei W, Zhu S, Zhu J, Shi Y, Lin T, Hao E, Hayek A, Deng H, Ding S (2009) Generation of rat and human induced pluripotent stem cells by combining genetic reprogramming and chemical inhibitors. Cell Stem Cell 4(1):16–19.
11. Li W, Ding S (2010) Generation of novel rat and human pluripotent stem cells by reprogramming and chemical approaches. Methods Mol Biol 636:293–300
12. Zhou H, Li W, Zhu S, Joo JY, Do JT, Xiong W, Kim JB, Zhang K, Scholer HR, Ding S (2010) Conversion of mouse epiblast stem cells to an earlier pluripotency state by small molecules. J Biol Chem 285:29676–29680
13. Hayashi K, Lopes SM, Tang F, Surani MA (2008) Dynamic equilibrium and heterogeneity of mouse pluripotent stem cells with distinct functional and epigenetic states. Cell Stem Cell 3(4):391–401
14. Imamura M, Miura K, Iwabuchi K, Ichisaka T, Nakagawa M, Lee J, Kanatsu-Shinohara M, Shinohara T, Yamanaka S (2006) Transcriptional repression and DNA hypermethylation of a small set of ES cell marker genes in male germline stem cells. BMC Dev Biol 6(1):34

Chapter 4

Isolation and Culture of Rabbit Embryonic Stem Cells

Arata Honda

Abstract

Mammalian stem cells are invaluable research resources for the study of cell and embryonic development as well as practical tools for use in the production of genetically engineered animals and further therapeutics. It is important that we further our knowledge and understanding of a variety of stem cells from several different animal species before trials in humans commence. Here we describe methods for establishing rabbit embryonic stem (rES) cell lines with indefinite proliferation potential. rES cells attain maximum proliferation potential when cultured at a feeder cell density of one-sixth of that of full confluency. Higher and lower densities of feeder cells induced ES cell differentiation or division arrest. Fibroblast growth factor (FGF)2 can maintain the undifferentiated status of rES cells; however leukemia inhibitory factor (LIF) is dispensable. Under optimized conditions, rES cells could be passaged by trypsinization 50 times. This culture system enabled efficient gene transduction and clonal expansion from single cells. rES cells grew as flat monolayer cell colonies, as reported for monkey and human ES cells, and expressed pluripotency markers. Embryoid bodies and teratomas formed readily in vitro and in vivo, respectively. Characterization of ES cells from different species is important for establishing common features of pluripotency. We have demonstrated the similarity of ES cells between rabbit and humans. These cell lines could be applied directly using gene-targeting techniques, or in combination with induced pluripotent stem cells. Thus, rES cells are a suitable model for studying human transplantation therapy and disease treatments.

Key words Rabbit, Embryonic stem cells, Pluripotent stem cells, Feeder cells, Fibroblast growth factor 2

1 Introduction

Embryonic stem (ES) cells are pluripotent and can self-renew indefinitely, and thus are potential donors for cell transplantation therapy and as source material for the production of genetically modified animals [1–3]. Although ES cell lines have been derived from mice and humans and are used extensively, the development of such lines from other mammals requires further technical improvement. In pigs, although the generation of porcine embryonic stem cell lines was first published in 1990, no conclusive results have been obtained from the isolation and propagation of putative pESC lines [4–7].

Ramiro Alberio (ed.), *Epiblast Stem Cells: Methods and Protocols*, Methods in Molecular Biology, vol. 1074,
DOI 10.1007/978-1-62703-628-3_4,

The laboratory rabbit (*Oryctolagus cuniculus*) has long been used for developing new reproductive and stem cell-related technologies since the first embryo transfer experiments in 1897 [8]. Basic reproductive engineering techniques, including in vitro fertilization and embryo cryopreservation, have long been established [9–11]. Nowadays, more advanced techniques including transgenesis, nuclear transfer, and intracytoplasmic sperm injection have become available. Rabbits are easy to maintain and handle, and because they are larger than mice or rats enable us to perform surgical operations on any tissue or organ.

Several ES cell lines were developed and characterized as showing major important characteristics in common with human ES cells [12–16]. "The true pluripotent state (e.g., it can readily differentiate into three germ layers in vitro and in vivo) is of practical importance because ES cells are expected to differentiate properly into multiple cell types [17]." Although the most convincing evidence for this state is the generation of germ line-competent chimeric offspring, this experimental system does not work in species other than mice and rats [18–20]. Instead, in other species including rabbits, the generation of teratomas with confirmed formation of all three germ layers (endoderm, mesoderm, and ectoderm) has to be used as a validation system. Recent molecular analysis suggests that the latter type of pluripotent cells corresponds to mouse epiblast stem cells, which are established from epiblast cells of postimplantation embryos [21–23].

This chapter describes an efficient and reproducible technique for establishing rES cells. The critical feature of the protocol is the density of feeder cells, which determines the fate of rES cells. Stable rES cell lines can be derived and propagated by optimizing the density of feeder cells.

2 Materials

2.1 Superovulation

1. Porcine follicle-stimulating hormone (pFSH; Antrin-R, Kawasaki Pharmaceuticals, Inc., Kanagawa, Japan) is suspended in physiological saline at 0.5 IU/ml.
2. Human chorionic gonadotropin (hCG; GONATROPIN, ASKA Pharmaceutical Co., Tokyo, Japan) is suspended in physiological saline at 100 IU/ml.

2.2 Egg Recovery

1. *RD (HEPES) medium*: 50 % DMEM, 50 % RPMI 1640, 0.5 % GlutaMAX.
2. *RD medium*: 50 % DMEM, 50 % RPMI 1640, 4 mg/ml bovine serum albumin (*see* **Note 1**).

3. Mineral oil.
4. Glass capillaries (Drummond, Calibrated pipettes 2-000-100).

2.3 Dissecting Embryos

1. 30-gauge needles.
2. Glass capillaries (Drummond, Calibrated pipettes 2-000-100).

2.4 Initial Phase of rES Cell Derivation

1. 0.1 % gelatin.
2. Mitomycin C-treated primary mouse embryonic fibroblast feeder cells (PMEFs) (*see* **Note 2**).
3. *PMEF medium*: DMEM, 10 % fetal bovine serum, penicillin–streptomycin.
4. DMEM-F12.
5. 0.1 % trypsin–DMEM-F12.
6. *rES medium (rESM)* with murine leukemia inhibitory factor (*mLIF*): 78 % DMEM-F12, 20 % knockout serum replacement, 2 mM GlutaMAX, 1 % nonessential amino acids, 0.1 mM β-mercaptoethanol, 10^3 IU/ml mLIF (ESGRO, Gibco), 8 ng/ml human recombinant FGF2 (hr-bFGF).

2.5 Stable Phase of rES Cell Derivation

1. 0.1 % gelatin.
2. PMEF.
3. PMEF medium.
4. DMEM-F12.
5. 0.1 % trypsin–DMEM-F12.
6. *rESM*: 78 % DMEM-F12, 20 % knockout serum replacement, 2 mM GlutaMAX, 1 % nonessential amino acids, 0.1 mM β-mercaptoethanol, 8 ng/ml hr-bFGF.

2.6 Cryopreservation

1. DMEM-F12.
2. 0.1 % trypsin–DMEM-F12.
3. rESM.
4. Cell Banker 2 (Juji Field Inc., Tokyo, Japan).
5. BICELL (Nihon Freezer Co., Tokyo, Japan).

2.7 Recovery of Cryopreserved rES Cells

1. 0.1 % gelatin.
2. PMEF.
3. PMEF medium.
4. DMEM-F12.
5. 0.1 % trypsin–DMEM-F12.
6. rESM.

3 Methods

3.1 Superovulation

1. Mature female-specific pathogen-free rabbits (Japanese White) weighing 4 to 6 kg are housed individually, fed a commercial pelleted diet, and kept at 24 °C. Female rabbits are induced to superovulate by six serial subcutaneous injections of 3 IU of pFSH dissolved in saline given at a dose of 0.5 IU/ml every 12 h for 3 days. The females are then given 100 IU of human chorionic gonadotropin intravenously in saline 72 h after the start of superovulation. Each female is mated with two fertile male rabbits immediately after injection of hCG.

3.2 Egg Recovery

1. Glass capillaries are pulled by hand over a very small flame to a diameter of 150–200 μm.
2. The superovulated female rabbits are euthanized with an overdose of anesthesia 16–18 h after hCG injection. The oviduct ampulla are recovered and flushed with 10 ml of RD (HEPES) medium, and the recovered 1–2 cell embryos are kept in RD medium supplemented with 4 mg/ml FBS at 38 °C in 5 % CO_2–5 % O_2 in air. To develop to the blastocyst, the recovered embryos are incubated in RD medium at 38 °C in 5 % CO_2–5 % O_2 in air for 48–60 h after the start of in vitro culture (*see* **Note 1**).

3.3 ES Cell Derivation (Preparation)

1. 12 h before dissecting the embryos, the PMEFs are prepared by seeding mitomycin C-treated PMEFs onto gelatinized four-well plates at a concentration of feeder cells of $36 \times 10^3/cm^2$ (Fig. 1) using routine methods for mouse ES cells.
2. Several hours before dissecting the embryos, the PMEF medium is aspirated from the four-well plates, the wells are

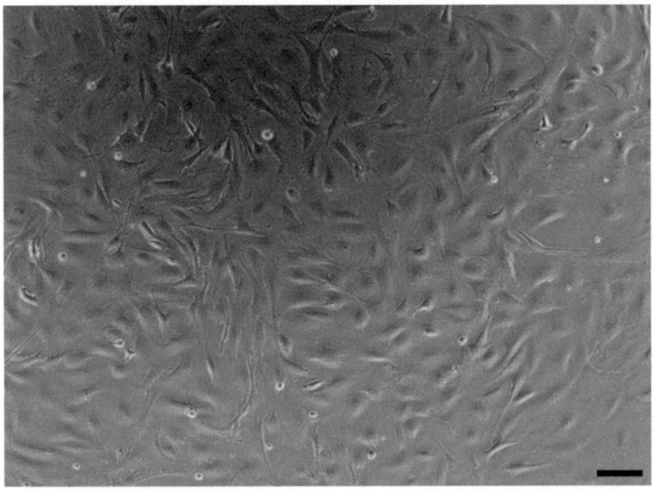

Fig. 1 Representative PMEF density of $36 \times 10^3/cm^2$. Scale bar: 100 μm

washed with DMEM-F12, and the medium is replaced with 0.5 ml of rESM (with mLIF). The plates are returned to the incubator and allowed to equilibrate.

3. 1 h before dissecting the embryos, a 6 cm dish is prepared with four drops (50 μl/drop) of RD (HEPES) medium, and covered with mineral oil.
4. 1 h before dissecting the embryos, a 6 cm dish is prepared with 10 drops (16 μl/drop) of RD medium, and covered with mineral oil. The plates are returned to the incubator and allowed to equilibrate.
5. Glass capillaries are pulled by hand over a very small flame to a diameter of 150–200 μm.

3.4 ES Cell Derivation (Dissecting Embryos)

1. Four embryos are transferred gently through the one drop (16 μl) on the collection dish.
2. The rabbit blastocyst zona pellucida is dissected using two 30-gauge needles as shown in Fig. 2 (*see* **Note 3**).
3. The blastocyst is transferred to an individual well containing PMEF (*see* **Note 2**). The plates are returned to incubator and left undisturbed for 4–7 days.

3.5 ES Cell Derivation (Primary Disaggregation of Embryo Explants)

On day 7–10 of incubation, the inner cell mass (ICM) explants are prepared for disaggregation. Each disaggregation is done individually.

1. 12 h before disaggregating ICM explants, PMEFs are prepared by seeding mitomycin C-treated PMEFs onto gelatinized four-well plates at a concentration of feeder cells as $6 \times 10^3/\text{cm}^2$, which is one-sixth of the density at full confluency (Fig. 3).
2. Several hours before dissecting the embryos, the PMEF medium is aspirated from the four-well plates, the wells are washed with DMEM-F12, and the medium is then replaced with 0.5 ml of rESM (with mLIF). The plates are returned to the incubator and allowed to equilibrate.
3. Glass capillaries are pulled by hand over a very small flame to a diameter of 150–200 μm.
4. The outgrowth of the ICM is scratched using a glass capillary to disaggregate some of the cell clumps (Fig. 4).
5. The clumps of cells are transferred to the receiving well and distributed gently over the monolayer of PMEFs. The plate is returned to the incubator.

3.6 rES Cell Derivation (Secondary Passaging of rES Cells from Primary Disaggregation)

1. If ≤4 colonies are generated 4–7 days after primary disaggregation, the primary disaggregation procedure is repeated.
2. If ≥5 colonies are generated, mitomycin C-treated PMEFs are prepared in 12-well plates at a concentration of feeder cells of

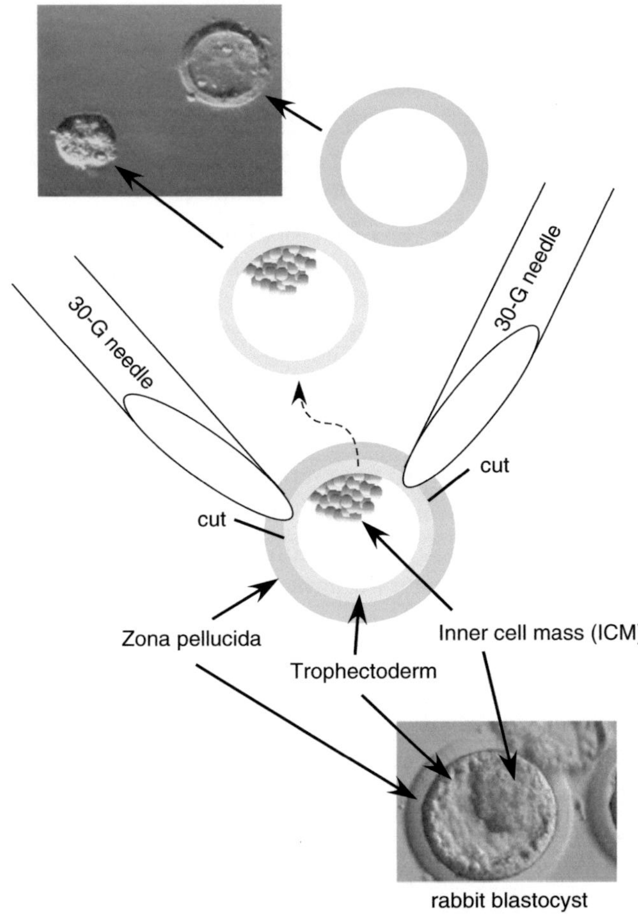

Fig. 2 Dissection of the zona pellucida using 30-gauge needles

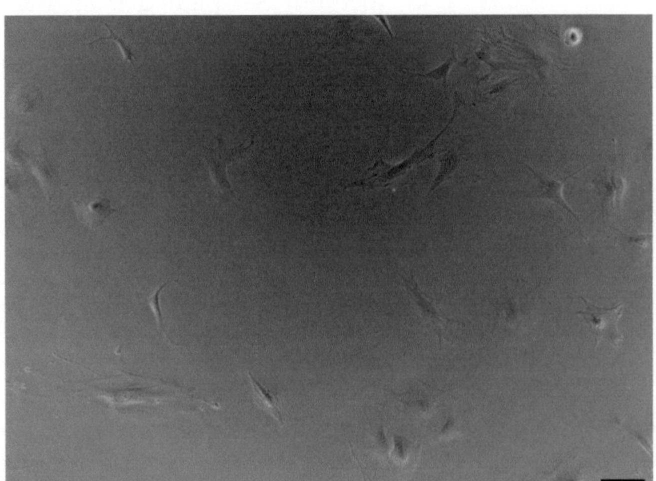

Fig. 3 Representative PMEF density of $6 \times 10^3/cm^2$. Scale bar: 100 μm

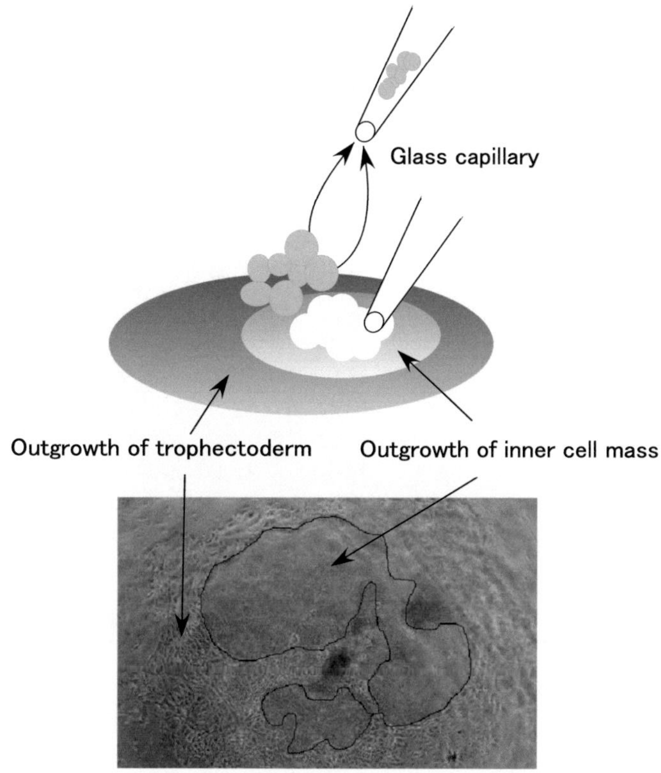

Fig. 4 Primary disaggregation of ICM explants using a glass capillary

$6 \times 10^3/cm^2$. The colonies are incubated overnight, and the medium is replaced with rESM.

3. The ES cell colonies are washed with DMEM-F12.
4. The cells are immersed in 0.3 ml of trypsin solution (0.1 % trypsin in DMEM-F12), the trypsin is discarded immediately, and the cells are incubated for 1 min at room temperature (*see* **Note 4**).
5. One-half milliliter of rESM is added, and the cell clumps are disaggregated with a blue tip (P-1000) to produce single cells. The entire content of the well is transferred to a 15 ml centrifuge tube.
6. The tube is centrifuged at $120 \times g$ for 3 min at room temperature.
7. The supernatant is discarded, and the cell pellet is resuspended in rESM. The entire content of the tube is transferred to a fresh 12-well PMEF plate. The next day, the medium is aspirated and replaced with rESM. Further passaging is performed on PMEFs at a concentration of $6 \times 10^3/cm^2$.

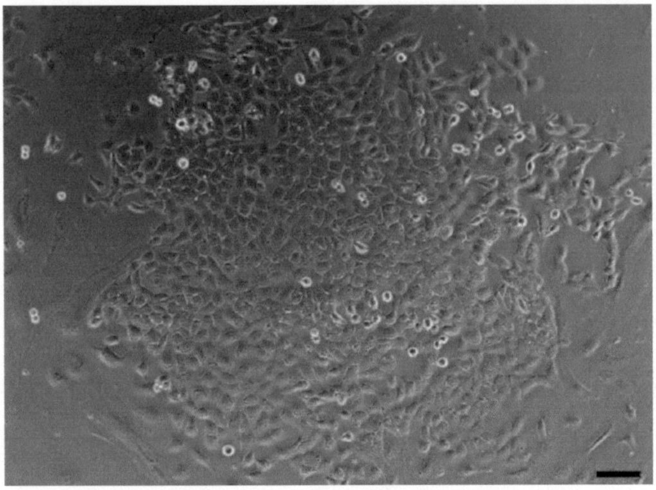

Fig. 5 Typical morphology of an rES cell colony. Scale bar: 100 μm

3.7 rES Cell Derivation (Initial Phase of Derivation)

In the initial phase of derivation (passage numbers 3–10), rES cells proliferate moderately. These cells grow as monolayer colonies with a doubling time of 13–18 h. However, after several passages, the cell cycle accelerates to 11–13 h. To prevent an unexpected acceleration, rES cells should be seeded at different densities to produce various cell numbers using 2–3 plots. It takes 4–5 days to grow to confluence from 3 to 5×10^3 cells per well (six-well plate). Cells are fed daily thereafter with rESM until confluent. After acceleration of the cell cycle (passages 5–15), rESM with mLIF is replaced by rESM without mLIF.

3.8 rES Cell Derivation (Stable Phase of Derivation)

In rES cells exhibiting stable proliferation (Fig. 5), 3–4 days are required for the cells to grow to confluence from 3 to 5×10^3 cells per well in a six-well plate. Cells are fed daily thereafter with rESM until confluent and then trypsinized using 0.1 % trypsin–DMEM-F12 for all passaging and cryopreservation steps.

1. rES cells are washed with DMEM-F12.
2. Cells are immersed in 1 ml of trypsin solution (0.1 % trypsin–DMEM-F12), the trypsin is discarded immediately, and the cells are incubated for 1 min at room temperature.
3. 1 ml of rESM is added, and the cell clumps are disaggregated with a blue tip (P-1000) to produce single cells. The entire content of the well is transferred to a 15 ml centrifuge tube.
4. The tube is centrifuged at $120 \times g$ for 3 min at room temperature.
5. The supernatant is discarded, and the cell pellet is dissolved in 1 ml of rESM. The cells are counted, and an appropriate number ($3\text{–}5 \times 10^3$ cells per well in a six-well plate) of cells are seeded onto a PMEF dish.

3.9 Cryopreservation

For freezing, 4–5 vials of cells are created from a confluent 6 cm dish.

1. 2–3 h before cryopreservation, the medium is replaced with fresh rESM.
2. The ES cells are washed with DMEM-F12.
3. The cells are immersed in 1 ml of trypsin solution (0.1 % trypsin–DMEM-F12), and the trypsin is discarded immediately. The cells are then incubated for 1 min at room temperature.
4. 3 ml of rESM is added, and the cells clumps are disaggregated with a blue tip (P-1000) to produce single cells. The entire content of the well is transferred to a 15 ml centrifuge tube.
5. The tube is centrifuged at 120×*g* for 3 min at room temperature.
6. The supernatant is discarded, and the cell pellet is dissolved in prechilled (4 °C) Cell Banker 2 (500 μl/vial).
7. The vials are transferred to a cell container (e.g., BICELL) and frozen at –80 °C for more than 3 h.
8. The vials are transferred to liquid nitrogen.

3.10 Recovery of Cryopreserved rES Cells

1. The vial is thawed rapidly in a 42 °C water bath, and 5 ml of rESM is added to the vial.
2. The cells are recovered by centrifugation at 120×*g* for 3 min at room temperature and plated onto PMEFs with rESM (*see* **Note 5**).
3. The next day, the old medium is replaced with fresh rESM, and the cells are fed daily thereafter until confluent.
4. Once confluent, the cells are passaged as indicated above in Subheading 3.8, **steps 1** and **5**.

4 Notes

1. RD medium should be prepared using fresh basal medium.
2. For primary culture of ICM explants, PMEFs are seeded onto four-well culture plates at a concentration of $36 \times 10^3/\text{cm}^2$.
3. In preliminary experiments, the mucin coat and zona pellucida of rabbit blastocysts flushed from the uteri of naturally mated females on postcoitus day 3.5 were removed either chemically using acidic Tyrode's solution (pH 2.4) or enzymatically using pronase (0.5 % in PBS). However, because of damage to blastocysts, it was not possible to establish rES cell lines. We strongly recommend removing the blastocyst from the zona pellucida by mechanical dissecting using a 30-gauge needle or artificial zona shedding of a blastocyst using a Piezo micromanipulator [14].

4. Although the rES cell displays features typical of epiblast stem cells, it can be clonally expanded by trypsin treatment in the absence of the ROCK inhibitor Y27632 [14, 16].
5. To adjust cell growth appropriately, the recovered rES cells should be seeded using a different proportion of each vial in 2–3 wells (e.g., one-third and two-thirds of the vial content to a well, respectively).

Acknowledgments

This work was supported by PRESTO of the Japan Science and Technology Agency with a Grant-in-Aid for Scientific Research (B).

References

1. Wang B, Zhou J (2003) Specific genetic modifications of domestic animals by gene targeting and animal cloning. Reprod Biol Endocrinol 1:103
2. Downing G, Battey JF Jr (2004) Technical assessment of the first 20 years of research using mouse embryonic stem cell lines. Stem Cells 22:1168–1180
3. Shufaro Y, Reubinoff BE (2004) Therapeutic applications of embryonic stem cells. Best Pract Res Clin Obstet Gynaecol 18:909–927
4. Evans MJ et al (1990) Derivation and prelimi nary characterization of pluripotent cell lines from porcine and bovine blastocysts. Theriogenology 33:125–128
5. Piedrahita JA, Anderson GB, Bondurant RH (1990) Influence of feeder layer type on the efficiency of isolation of porcine embryo-derived cell-lines. Theriogenology 34:865–877
6. Strojek RM (1990) A method for cultivating morphologically undifferentiated embryonic stem-cells from porcine blastocysts. Theriogenology 33:901–913
7. Vackova I, Ungrova A, Lopes F (2007) Putative embryonic stem cell lines from pig embryos. J Reprod Dev 53:1137–1149
8. Biggers JD (1991) Walter Heape, FRS: a pioneer in reproductive biology. Centenary of his embryo transfer experiments. J Reprod Fertil 93:173–186
9. Chang MC (1959) Fertilization of rabbit ova in vitro. Nature 184:466–467
10. Kasai M et al (1992) High survival of rabbit morulae after vitrification in an ethylene glycol-based solution by a simple method. Biol Reprod 46:1042–1046
11. Liu JL et al (2004) Freeze-dried sperm fertilization leads to full-term development in rabbits. Biol Reprod 70:1776–1781
12. Graves KH, Moreadith RW (1993) Derivation and characterization of putative pluripotential embryonic stem cells from preimplantation rabbit embryos. Mol Reprod Dev 36:424–433
13. Fang ZF et al (2006) Rabbit embryonic stem cell lines derived from fertilized, parthenogenetic or somatic cell nuclear transfer embryos. Exp Cell Res 312:3669–3682
14. Honda A et al (2008) Stable embryonic stem cell lines in rabbits: potential small animal models for human research. Reprod Biomed Online 17:706–715
15. Intawicha P et al (2009) Characterization of embryonic stem cell lines derived from New Zealand white rabbit embryos. Cloning Stem Cells 11:27–38
16. Honda A, Hirose M, Ogura A (2009) Basic FGF and Activin/Nodal but not LIF signaling sustain undifferentiated status of rabbit embryonic stem cells. Exp Cell Res 315:2033–2042
17. Ying QL et al (2008) The ground state of embryonic stem cell self-renewal. Nature 453:519–523
18. Okita K, Ichisaka T, Yamanaka S (2007) Generation of germline-competent induced pluripotent stem cells. Nature 448:313–317
19. Buehr M et al (2008) Capture of authentic embryonic stem cells from rat blastocysts. Cell 135:1287–1298

20. Li P et al (2008) Germline competent embryonic stem cells derived from rat blastocysts. Cell 135:1299–1310
21. Brons IG et al (2007) Derivation of pluripotent epiblast stem cells from mammalian embryos. Nature 448:191–195
22. Tesar PJ et al (2007) New cell lines from mouse epiblast share defining features with human embryonic stem cells. Nature 448: 196–199
23. Nichols J, Smith A (2009) Naive and primed pluripotent states. Cell Stem Cell 4:487–492

Chapter 5

Goat Embryonic Stem-Like Cell Derivation and Characterization

Esmail Behboodi, Louise Lam, William G. Gavin, Alla Bondareva, and Ina Dobrinski

Abstract

Embryonic stem (ES) cells are derived from the inner cell masses of preimplantation embryos. ES cells are pluripotent cells with the capacity for long-term propagation and broad differentiation plasticity. These cells have an exceptional functional feature in that they can differentiate into all tissues and organs, including germ cells. Established ES cell lines have been generated in mouse, human, and nonhuman primate but derivation of ES cells in farm animals has been problematic. Several ES-like cell lines from farm animals have been reported to exhibit properties of pluripotency in vitro. However, only a few of them morphologically resemble ES cells, or express markers that are associated with established ES cell lines from mouse and humans. Methods for derivation, propagation, and differentiation of ES cells from domestic animals have not been fully established. In this chapter, we describe methods for isolation of goat ES (gES) cell lines from in vivo-derived blastocysts and characterization of markers indicative of pluripotency. In addition, we outline differentiation of gES cells into all three germ layers in vivo by forming teratomas as a hallmark of pluripotency.

Key words Embryonic stem cells, Goat, Teratoma, In vitro differentiation

1 Introduction

Stem cells are characterized by their ability for self-renewal and their capability to differentiate into a broad spectrum of cell types. Embryonic stem (ES) cell lines have been established in the mouse, nonhuman primate, and humans [1–3]. However, generation of ES cells from embryos of domestic animals has been more challenging. Recently, there have been reports on ES-like cell lines derived from bovine embryos [4–6]. Bovine ES-like cells have been derived either from 16-cell [7] or from day-7–9 blastocyst-stage embryos [7, 8]. These cell lines displayed high nucleus-to-cytoplasm ratio, competence for extended culture in vitro, and capacity to differentiate into embryoid bodies (EB), similar to mouse ES cells. A panel of markers indicative of pluripotency of ES cells has been

Ramiro Alberio (ed.), *Epiblast Stem Cells: Methods and Protocols*, Methods in Molecular Biology, vol. 1074,
DOI 10.1007/978-1-62703-628-3_5,

investigated in mouse, human, and bovine cells. Mouse and human ES cells display alkaline phosphatase activity whereas staining of bovine ES-like cells for this marker is more variable [7, 8]. The transcription factor Oct-4 (also known as Pou5f1), a key regulator in undifferentiated pluripotent cells, appears to be a primary transcription factor in ES cells. The presence of Oct-4 in bovine 16-cell embryos, inner cell mass (ICM), and goat blastocysts has been verified previously [9–11]. Nanog has also been previously reported in preimplantation-stage goat embryos [12]. Stage-specific embryonic antigen (SSEA)-1 is expressed in mouse ES cells while human ES-like cells possess SSEA-3 and SSEA-4, but not SSEA-1 antigens. Conflicting results have been reported for SSEA-3 and SSEA-4 in bovine ES cells. Although reports of the development of ES-like lines have been published in many species, evidence is lacking for pluripotency as determined by teratoma formation or germline transmission in vivo. Lately, there have been several reports on derivation of goat ES (gES) cells [11–15].

Here we describe the procedures for isolation of gES cells from in vivo-derived blastocysts and expression of markers indicative of pluripotency in the gES cells. We demonstrated the capacity of gES cells to differentiate into all three germ layers in vivo by forming teratoma as a hallmark of pluripotency [12].

2 Materials

2.1 Chemicals

1. Mitomycin C.
   (a) Add 2 ml PBS per vial (2 mg) to dissolve the powder.
   (b) Remove all the mitomycin-C solution from the vial with a 3 cc syringe.
   (c) Filter the solution with a 0.22 μm filter into a 15 ml tube.
   (d) The solution should be covered with aluminum foil (mitomycin-C is light sensitive).
   (e) Store at 4 °C and use within 2–4 weeks.
2. β-Mercaptoethanol.
3. Recombinant hLIF.
4. Basic fibroblast growth factor.
5. 0.25 % trypsin/EDTA.
6. DMEM medium.
7. Penicillin/streptomycin (P/S).
8. Medium 199.
9. Fetal bovine serum (FBS).

3 Methods

3.1 Superovulation and Embryo Collection

The method requires close collaboration with a veterinarian who will supervise the goat breeding and surgery. On average, goats produce five pre-implantation stage embryos per donor.

1. Prime donors hormonally according to established methods [16, 17].
2. Synchronize estrus in fertile (multiparous) donors between approximately 1–5 years of age for best results; the optimal timing for this procedure is between October and late January in the Northern hemisphere.
3. Use intravaginal sponges (or subcutaneous implants) containing 60 mg medroxyprogesterone acetate (Veramix, Pharmacia & Upjohn, Mississauga, ON, Canada) for 10 days with an injection of 125 μg cloprostenol (Estrumate, Schering, Pointe-Clair, QC, Canada) or 5–10 mg of Lutalyse (PGF2a, Pfizer) at sponge removal.
4. Induce superovulation with a total equivalent of 160 mg FSH (Folltropin-V, Bioniche, Belleville, ON, Canada) administered twice daily at decreasing doses over 4 days starting 48 h prior to sponge removal [16].
5. 100 μg GnRH (Factrel, Pfizer Canada, Kirkland, QC, Canada) is given 36 h after sponge removal followed by artificial insemination with fresh semen 12 and 24 h later.
6. Collect embryos by surgical uterine flushing under general anesthesia 6 days after breeding to collect blastocyst-stage embryos.
7. Use a retrograde uterine flush by injecting approximately 30 ml of EmCare medium (ICPBio Reproduction, Auckland, New Zealand) with a 18–21 G catheter through the oviductal opening and recovering the fluid through a 8–12 Fr. Foley catheter placed at the base of each uterine horn.

3.2 Harvesting Goat Fetal Fibroblast Cells

1. For harvesting fetal fibroblasts, surgically remove a fetus at day 30–35 post breeding and place it in phosphate-buffered saline (PBS).
2. Transfer the fetus to a 10 cm Petri dish with 10 ml DMEM medium supplemented with P/S.
3. Hold the head of the embryo with blunt forceps, and carefully remove all the viscera using sharp forceps.
4. Cut off the limbs and tail of the embryo using the forceps. Finally, remove the head of the embryo and transfer the remaining trunk to a new Petri dish with PBS supplemented with P/S.

5. Mince the eviscerated embryo sections very well with two razor blades and transfer the pieces to 15 ml tube containing DMEM medium supplemented with 15 % FBS. Pellet the small pieces of tissue by centrifugation (100 × *g* for 5 min).
6. Culture fetal tissues in 125 cm^2 flasks in DMEM medium supplemented with 15 % FBS for 4–5 days.
7. Harvest monolayer of cells by trypsinization.

3.3 Cryopresevation of Goat Fetal Fibroblast Cells

1. Warm the necessary volume of trypsin in water bath at 37 °C for at least 5 min.
2. Aspirate media from the monolayer cell culture plate.
3. Add the appropriate amount of trypsin (0.5 ml/well in a 12-well plate and 2 ml/well in or 6-well plate).
4. Incubate at 37 °C with 5 % CO_2 for 5 min. It is important to dissociate colonies into a single-cell suspension.

1. Add an equal amount of DMEM medium supplemented with 15 % FBS to stop the trypsinization process.
2. Centrifuge at 500 × *g* for 5 min.
3. Aspirate the medium and resuspend the pellet in 1 ml DMEM medium supplemented with 15 % FBS.
4. Take a 10 μl aliquot to do a cell count using a hemocytometer.
5. Place 1 million cells in 0.5 ml of culture medium to each cryovial. Add 0.3 ml of FBS and 0.2 ml of dimethyl sulfoxide (DMSO) drop-wise to the cells (final concentration of FBS and DMSO is 30 and 20 %, respectively).
6. Rapidly transfer the cryovials into a freezing container (e.g., Nalgene® Mr. Frosty) and place it to −80 °C for a minimum of 24 h.
7. Transfer the cryovials to long-term liquid nitrogen storage.

3.4 Deactivation of Goat Fetal Fibroblast Cells as Feeder

1. Remove the goat fetal fibroblasts (GFF) from liquid nitrogen and thaw immediately in a 37 °C water bath.
2. Add the cell suspension into 5–10 ml DMEM medium supplemented with 15 % FBS in a 15 ml tube and mix.
3. Centrifuge the cell suspension at 500 × *g* for 5 min.
4. Aspirate the supernatant and resuspend the pellet in 2–4 ml DMEM medium supplemented with 15 % FBS. Add cells to a T-75 flask containing 15 ml DMEM medium.
5. Culture at 37 °C until cells reach confluence (2–3 days).
6. Aspirate medium from flask and add 20 ml of DMEM medium with mitomycin-C (10 μg/ml).
7. Leave mitomycin-C solution in flask for 1.5–2 h in incubator.
8. Aspirate medium from flask.

9. Wash flasks twice with 20 ml of warm PBS.
10. Add 2–5 ml 0.25 % trypsin/EDTA to each flask and leave in incubator for 1–2 min at 37 °C. Tap the flask gently to detach the cells.
11. Add 10 ml of DMEM medium to flask and pipette gently up and down to further disaggregate the cells.
12. Centrifuge the cells at $500\times g$ for 5 min and discard supernatant.
13. Resuspend cells in appropriate volume of DMEM supplemented with 15 % FBS. Plate cells at 6×10^4 cells/cm^2 for culture to be used as feeder cells.
14. Deactivated goat fibroblast cells can be used up to 7–10 days as feeder.

3.5 Goat ES Cell Collection

1. Culture day-6–8 in vivo-derived embryos on GFF feeder layers in gES cell medium consisting of DMEM medium supplemented with 0.1 mM 2-mercaptoethanol, 0.1 mM MEM non-essential amino acids, 2 mM L-glutamine hLIF (1000 Uml) EGF, FGF and 15 % FBS in a 12-well plate. Floating embryos will expand and a few expanded ICM cells may form a disc-like structure. Some embryos will attach to the feeder cells and have an expanded ICM with surrounding trophoblastic cells.
2. There are two main methods to derive cell lines from blastocysts.
   (a) Culture of blastocyst on feeder and isolation of ICM (Fig. 1 top):
      - Remove typical primary colonies of ES-like cells, the proliferating ICM cells that are surrounded by trophoblastic cells, form 4–5 days after blastocyst attachment to feeder cells.
      - Culture the isolated outgrowth of ICM on new goat feeder cells.

   (b) Epiblast/embryonic disc isolation and culture on feeder (Fig. 1 bottom):
      - Isolate embryonic discs mechanically on day 10–12 post fertilization and culture on goat feeder cells in DMEM medium containing 1,000 U/ml hLIF and 15 % FBS (floating embryos form structures resembling ICM discs surrounded by trophoblastic cells 3–4 days in culture).
      - The embryonic discs outgrow into colonies on days 4–8 in culture. Isolate compact colonies (70–120 μm) of cells from the outgrowths and mechanically passage

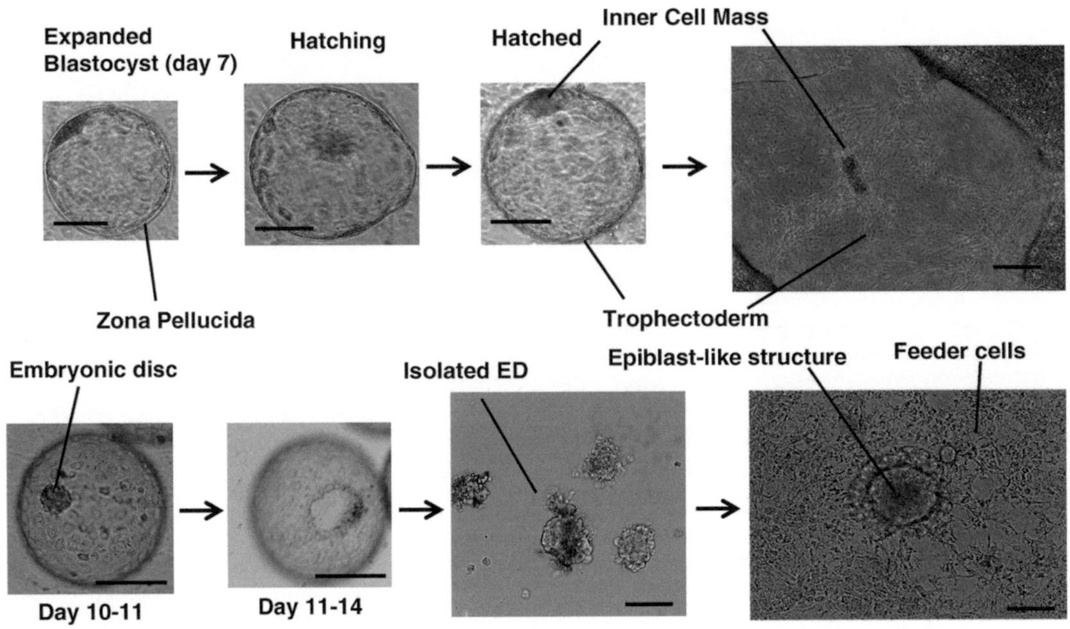

Fig. 1 Caprine embryos at different developmental stages and derivation of ES cells. *Top*: Flowchart depicting day-7 blastocyst hatching from zona pellucida. The hatched blastocyst is then cultured on feeder cells. Blastocysts ~400 μm in diameter. Bar: 100 μm. Bottom: The well-defined embryonic disc of a day-12 goat embryo surrounded by trophoblastic cells is isolated mechanically and cultured on feeder cells. Day-10–11 goat embryo having a well-defined inner cell mass (ICM). Day-11–14 blastocysts have a clear embryonic disc that is mechanically isolated from trophoblastic cells. Embryonic discs (ED) are cultured on feeder cells. Blastocysts ~700 μm in diameter. Bars: 100 μm (modified from ref. 12)

onto fresh goat feeder cells (passage 1, plate on a 200 mm^2 gelatinized growth surface).

3. Collect colonies mechanically under the microscope, disaggregate by pipetting, and seed on feeder cells in 6-well culture plates for expansion after passages 4–5 (Fig. 2a–e).

3.6 Goat ES Cell Passaging

1. Gelatinize plate 2 h before passaging the cells.
2. Pre-warm 0.25 % trypsin/EDTA at 37 °C for at least 5 min.
3. Aspirate media from the gES culture plate. Add trypsin (0.5 ml/well for a 12-well plate and 2 ml/well for or 6-well plate).
4. Incubate at 37 °C with 5 % CO_2 for 3–5 min. The cells are ready when the edges of the colony are rounded up and curled away from the feeder cells on the plate. Stop the reaction by adding an equal volume of DMEM medium containing 15 % FBS.
5. Using a 5 ml pipette, scrape and rinse the colonies off the plate with DMEM medium supplemented with 15 % FBS.
6. Transfer cell suspension to a 15 ml tube.

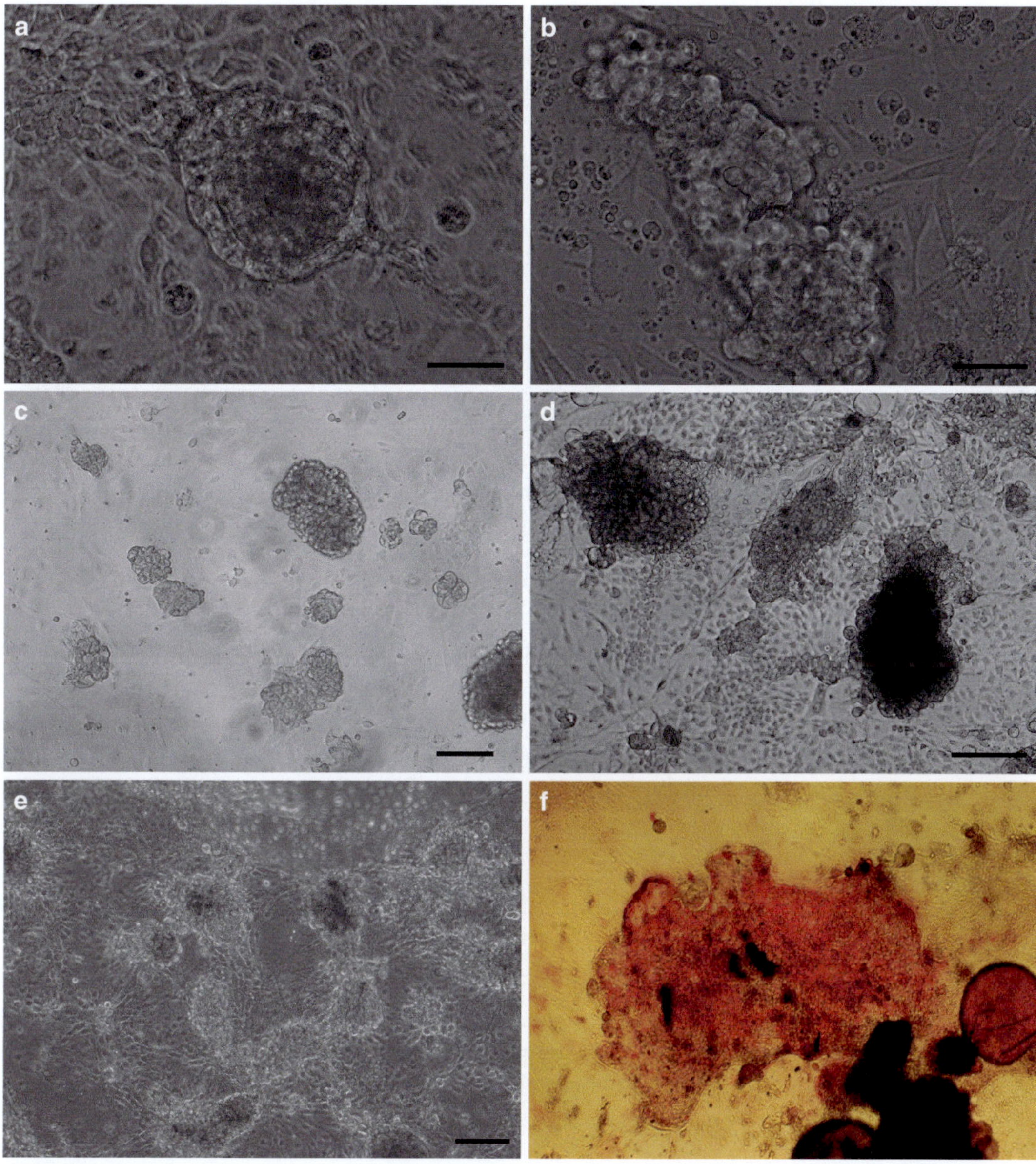

Fig. 2 Goat ES cells after culture. (**a**) Epiblast-like structure at day 5 post attachment onto goat feeder cells. (**b**) Enlargement of the epiblast-like structure 3 days post transfer to goat feeder cells. (**c**) 3 days post transfer to goat feeder cells. (**d**) Large colonies at day 5 post transfer to goat feeder cells. (**a**, **b**, **d**): Bars: 100 μm. (**c**) Bar: 60 μm. (**e**) Goat ES cell colonies at passage 4. Bar: 100 μm. (**f**) Alkaline phosphatase staining in goat ES cell colonies

7. Break up the colonies by pipetting up and down against the bottom of the tube until there appears to be a fine suspension of cells (no clumps of cells remaining).
8. Spin cells at $500 \times g$ for 5 min.

9. Resuspend cells in an appropriate volume of gES medium. Usually, gES cells are split 1:3–1:4 at every passage. Plate cells by adding the appropriate volume of cell suspension in each well.
10. To freeze gES cells, follow protocol as outlined in Subheading 2, **item 4**. Cryopreservation of GFF cells. Instead of complete dissociation of colonies into single-cell suspension, it is important to GENTLY pipette gES cells and leave 3–5 cell clusters (*see* **Note 1**).

3.7 Goat ES Cell Thawing

1. Gelatinize 6- or 12-well plates 24 h before thawing gES cells.
2. To avoid alkalinity change, place the growth medium (10 ml for one vial) in a 15 ml tube and place into incubator at 37 °C for at least 15 min to allow the medium to reach its normal pH (7.4).
3. Remove cryovials from liquid nitrogen.
4. Quickly thaw the vials in a 37 °C water bath (approximately 1 min) until just a small crystal of ice remains in the vial.
5. Gently pipette the cells up and down once and place them in a 15 ml tube.
6. Slowly add 9.5 ml of DMEM medium supplemented with 15 % FBS to the tube. While adding medium, gently mix the cells in the tube.
7. Centrifuge the cells at 500 × *g* for 5 min.
8. During centrifugation, prepare the culture plate by aspirating excess gelatin and washing once with PBS (without Ca^{2+}/Mg^{2+}) and then once with the gES medium.
9. Resuspend gES cells in 10 ml of gES media.
10. Seed 1 ml of the gES cell suspension in each well of a 12-well plate or 3 ml of the gES cell suspension in each well of 6-well plate. Incubate at 37 °C in 5 % CO_2.
11. Change the medium daily.

3.8 Goat ES Cell Characterization

3.8.1 Alkaline Phosphatase

1. Culture gES cells on feeder layers for 2–3 days prior to staining, until colonies are formed. The feeder cells will serve as negative control for the staining procedure.
2. Remove culture medium from the plates and wash cells with PBS. Fix gES cells with 90 % methanol and 10 % formaldehyde for 2 min. Wash fixed cells twice with TBST (25 mM Tris–HCl, 0.14 M NaCl, 2.7 mM KCl, 0.1 % Tween-20) and stain with Napthol/Fast Red Violet solution for 15 min at room temperature in the dark.
3. Wash cells with TBST in order to terminate the staining reaction and maintain stained cells in PBS.
4. Observe the cells under a light microscope (*see* Fig. 2f).

3.8.2 Goat ES Characterization Using Pluripotency Markers Oct-4, Nanog, SSEA-4, and Sox2 by Immunocytochemistry

1. Culture gES cells without feeder cells in gES medium supplemented with hLIF (1,000 U/ml) on chamber slides (1×10^3 cells per chamber) for 3–4 days.
2. Fix cells with 4 % paraformaldehyde for 30 min at room temperature.
3. For permeabilization, incubate slides with 0.2 % Triton X-100 for 20 min.
4. Wash slides once with PBS and incubate with 5 % donkey serum for blocking of nonspecific binding.
5. Incubate slides overnight at 4 °C with primary monoclonal antibodies against mouse SSEA-4, Oct-4, Nanog, or Sox2 (Abcam, Cambridge, MA, USA) at concentrations of 1:50, 1:200, 1:200, and 1:200, respectively. Prepare the antibodies in PBS with 5 % donkey serum to further decrease nonspecific binding.
6. Wash slides with PBS for 5 min, add appropriate secondary antibody at 1:400 dilution, and incubate for 1 h at room temperature.
7. Wash slides three times with PBS and add mounting solution containing DAPI for nuclear staining.
8. Observe the cell staining under fluorescence microscope (*see* Fig. 3).

3.9 RT-PCR Assay: Amplification of Oct-4, Nanog, Sox2, and Lin28

1. Wash cultured cells once with PBS and lyse directly on the plate.
2. Extract total RNA from cultured cells with RNeasy Mini kit followed by treatment with DNase I according to the manufacturer's protocol.
3. The RNA quality and quantity can be determined using a spectrophotometer.
4. For reverse transcription, use 2 μg of total RNA in a final volume of 25 μl reaction containing 0.5 μg of Oligo $d(T)_{12–18}$, RT buffer (1×), 10 mM dithiothreitol, 0.5 mM of dNTP, 5 U of RNase-inhibitor, and 10 U of SuperScript II Reverse transcriptase.
5. Carry out reverse transcription at 42 °C for 1 h.
6. Primer sequences (5′–3′), length of amplified products, and annealing temperatures are shown in Table 1.
7. PCR products are resolved in 1.5 % agarose gel and imaged (*see* Fig. 4a).

3.10 Karyotyping

1. Culture gES cells in DMEM medium supplemented with 15 % FBS and growth factors (LIF, EGF, and FGF) and incubate overnight in culture medium containing 0.05 g/ml colcemid KaryoMax (Life-Technology, Burlington, ON, Canada) (*see* **Note 2**).
2. Harvest the cells by trypsinization, wash twice in PBS, and resuspend in 0.56 % KCl for 20 min at room temperature.

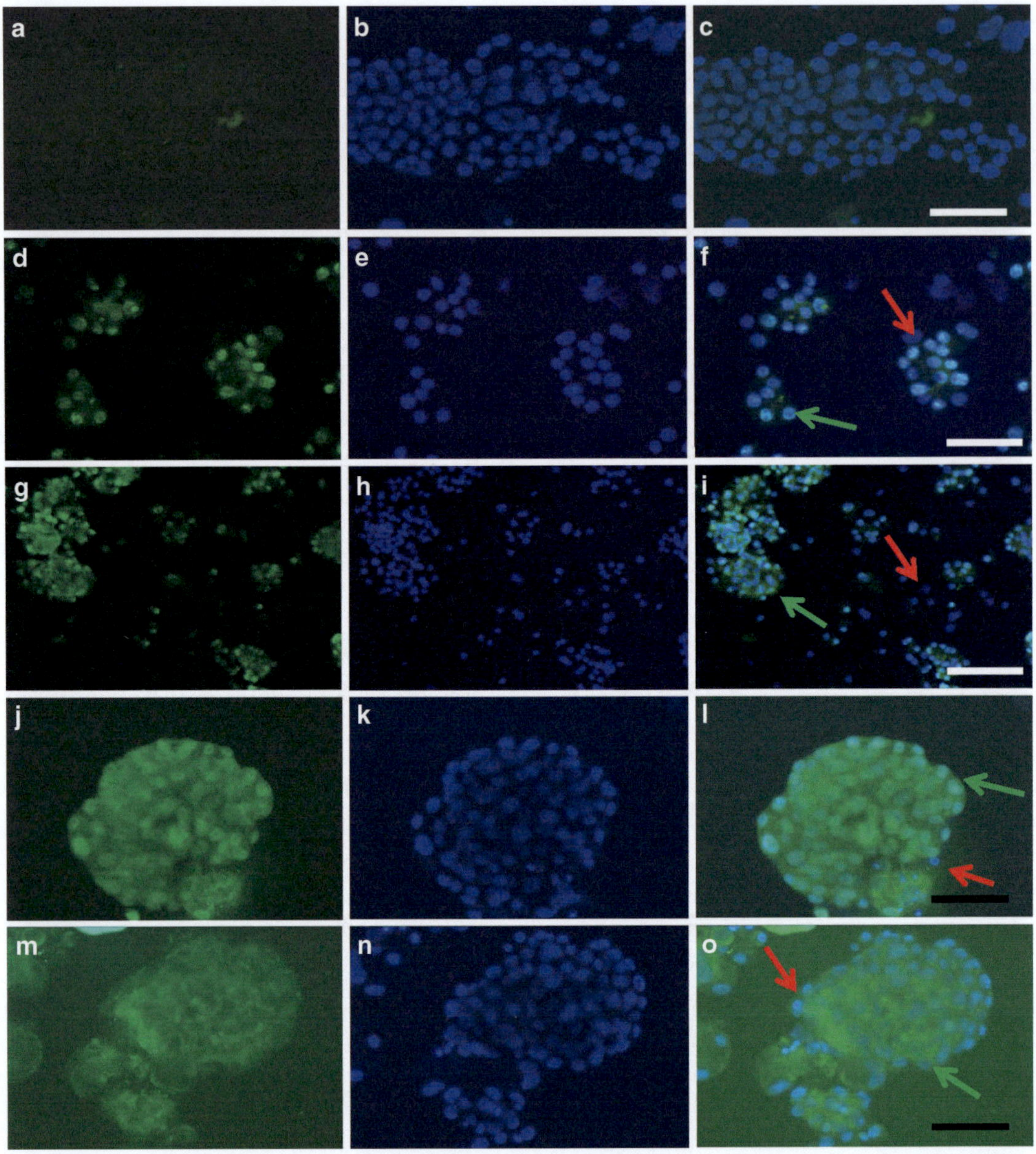

Fig. 3 Goat ES cells express markers of pluripotent cells. (**a**) Goat ES cell colonies with no primary antibody ((**b**) DAPI and (**c**) merged). (**d**) Goat ES cell colonies express Oct-4 ((**e**) DAPI and (**f**) merged); (**g**) Nanog ((**h**) DAPI and (**i**) merged); (**j**) Sox-2 ((**k**) DAPI and (**l**) merged); and (**m**) SSEA-4 ((**n**) DAPI and (**o**) merged). *Red arrows* point to examples of negative cells, *green arrows* point to examples of positive cells. (**a**, **b**, **c**, **j**, **k**, **l**, **m**, **n**, **o**): Bars = 100 μm. (**d**, **e**, **f**, **g**, **h**, **i**): Bar = 60 μm (modified from ref. 12)

3. Briefly, fix the cells in cold Carnoy's fixative for 30 min (3:1 methanol:glacial acetic acid), spread the cells on glass slides, and stain with Giemsa stain.
4. Analyze the metaphase spreads using a cytogenetic analysis system (*see* Fig. 4b).

Table 1
Primer sets for RT-PCR

Primers	Sequence, 5′–3′	Annealing T, °C	Length of the PCR fragment, bp
gPOU5F1(goat *Oct-4*), forward	AGGTGTTCAGCCAAACGACTATCTG	60.0	192
gPOU5F1(goat *Oct-4*), reverse	TCGGTTCTCGATACTTGTCCGCTT		
mPOU5F1 (mouse *Oct 4*), forward	AGGTGTTCAGCCAGACCACCATC	60.0	193
mPOU5F1 (mouse *Oct 4*), reverse	CTCGGTTCTCAATGCTAGTTCGCTT		
gNANOG, forward	AGGACAGCCCTGATTCTTCCACAAG	60.5	239
gNANOG, external reverse	TTCTCTGATTCTGGAACCAGGTCTTCAC		
gNANOG, internal reverse	GGAAAGTTCGTTCATTTGCTGGAGACTG	60.5	199
mNANOG, forward	AAGGCAGCCCTGATTCTTCTACCAG	60.5	198
mNANOG, reverse	GAAAGTTCTTGCATCTGCTGGAGGC		
gSox-2, external forward	TGCAGTACAACTCCATGACCAGCT	59.7	281
gSox-2, external reverse	GTAGTGCTGGGACATGTGAAGTCTG		
Sox-2 (goat and mouse), internal forward	CTACAGCATGTCCTACTCGCAGC	59.2	174
Sox-2 (goat and mouse), internal reverse	CGGGGAGGTACATGCTGATCAT		
gLin-28, external forward	CTGTAAGTGGTTCAACGTGCGCATG	61.0	480
gLin-28, external reverse	ATGGCAGGGCTGTGGATCTCTTC		
Lin-28 (goat and mouse), internal forward	GATGTCTTTGTGCACCAGAGTAAGCTG	60.5	226
Lin-28 (goat and mouse), internal reverse	TAGACCTCCACAGTTGTAGCACCTGTC		
GAPDH (goat and mouse), forward	TCATGACCACAGTCCATGCCATCACT	60.0	253
GAPDH (goat and mouse), reverse	GATGTCATCATATTTGGCAGGTTTCTCC		

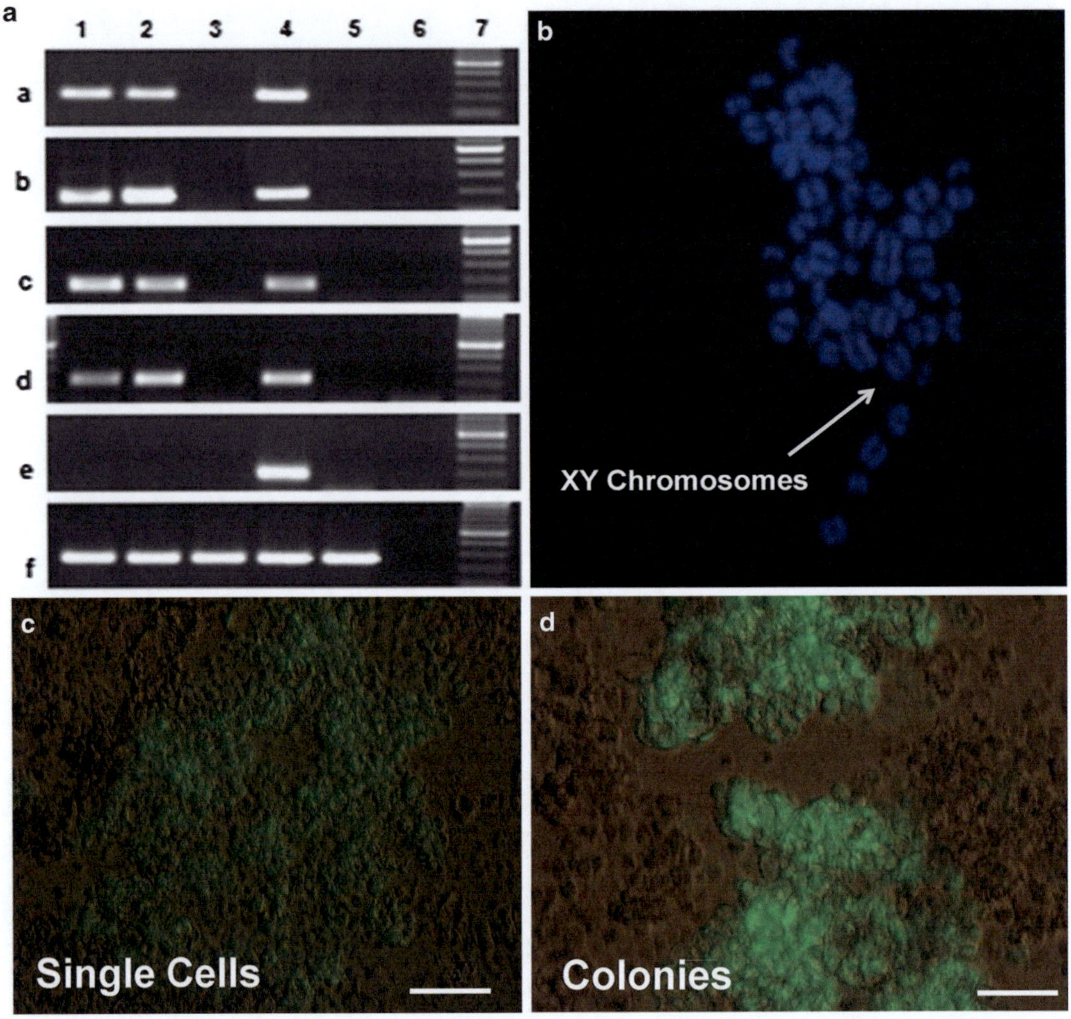

Fig. 4 Characterization of goat ES cells. (**a**) Gene expression in goat ES cells as assessed by RT-PCR: (*a*) *Oct-4*, expected size 192 bp; (*b*) *Nanog* with goat-specific and mouse-specific primers, respectively, expected size 199 bp; (*c*) *Sox-2*, expected size 174 bp; (*d*) *Lin-28*, expected size 226 bp; (*e*) *Nanog* with mouse-specific primers only, expected size 199 bp; (*f*) *GAPDH*, expected size 253 bp. *Lanes 1* and *2*, two different, representative preparations of gES cells; *3*, goat fetal fibroblasts (GFF); *4*, mouse ES cells (R1), positive control; *5*, mouse fetal fibroblasts (MFFs); *6*, no template control; *7*, 100 bp markers (from ref. 12). (**b**) Chromosome spread of goat ES cells at passage 60 (60, XY), (**c**) fluorescence microscopy image showing stable GFP expression in gES single cells and (**d**) in gES cell colonies. Bars = 100 μm

3.11 Goat ES Cell Transfection

1. Thaw frozen aliquots (5×10^5/ml) of gES cells.
2. Prepare the construct of interest (we selected CEeGFP plasmid) (*see* **Note 3**).
3. Introduce the construct into undifferentiated gES cells by lipid-mediated gene transfer (e.g., use lipofectamine (Life Technology, Burlington, ON, Canada) according to the manufacturer's instructions).

4. Derive stable clones by selection with G418 for 20 days and assess expression of the reporter gene by visualization of the fluorescent signal (*see* Fig. 4c).

3.12 In Vitro Differentiation of Goat ES Cells

1. Trypsinize colonies of undifferentiated gES cells and plate the cells at a concentration of 2×10^5 cells/cm^2 culture in DMEM medium supplemented with 15 % FBS and 2–3 μM/ml retinoic acid without LIF in 35 mm gelatin-coated dishes.
2. Incubate cells at 37 °C for 4–10 days and change medium every 48 h.
3. When structures resembling neurospheres appear floating in plates, rinse the plates 2–3 times with DMEM medium and harvest supernatant. Allow neurospheres to settle down in collection tube for 10–15 min in incubator, transfer them to gelatinized plates, culture for 3–4 days, and examine cells for neuronal and epithelial like morphology (Fig. 5).
4. To confirm differentiation into the neuronal lineage, use a polyclonal antibody raised in rabbits against human nestin (1:100, Abcam, Cambridge, USA) for immunohistochemical staining of cells.

3.13 Teratoma Formation

1. Aspirate medium from culture plates, and wash gES cells with PBS. Trypsinize gES colonies into single cells for 5 min, with gentle pipetting.
2. Collect the total pool of cells, centrifuge at $500 \times g$ for 5 min, aspirate medium, and wash cells with PBS by gently swirling the tube.
3. Determine the total number of cells and aliquot cells into tubes corresponding to the number of injection locations (e.g., six locations of injection: subcutaneous (left and right), intramuscular (left and right), testicular capsule (left and right)) and number of cells per location (e.g., five million cells without Matrigel (BD Biosciences, Mississauga, ON, Canada) per location, or 1–2 million cells with Matrigel per location).
4. Centrifuge cells in each tube at $500 \times g$ for 5 min. Aspirate supernatant and determine the approximate volume of the pellet for each tube. Each tube should contain the cells for each injection location.
   (a) Without Matrigel: Resuspend the cells in PBS (final volume depends on the number of injection locations) for each tube at a minimum concentration of 5 million cells/location.
   (b) With Matrigel: Resuspend the cells in PBS supplemented with chilled, liquid 30 % Matrigel (final volume depends on the number of injection locations) at a minimum concentration of one million cells/location. Mix contents by GENTLE pipetting and keep the cells on ice.

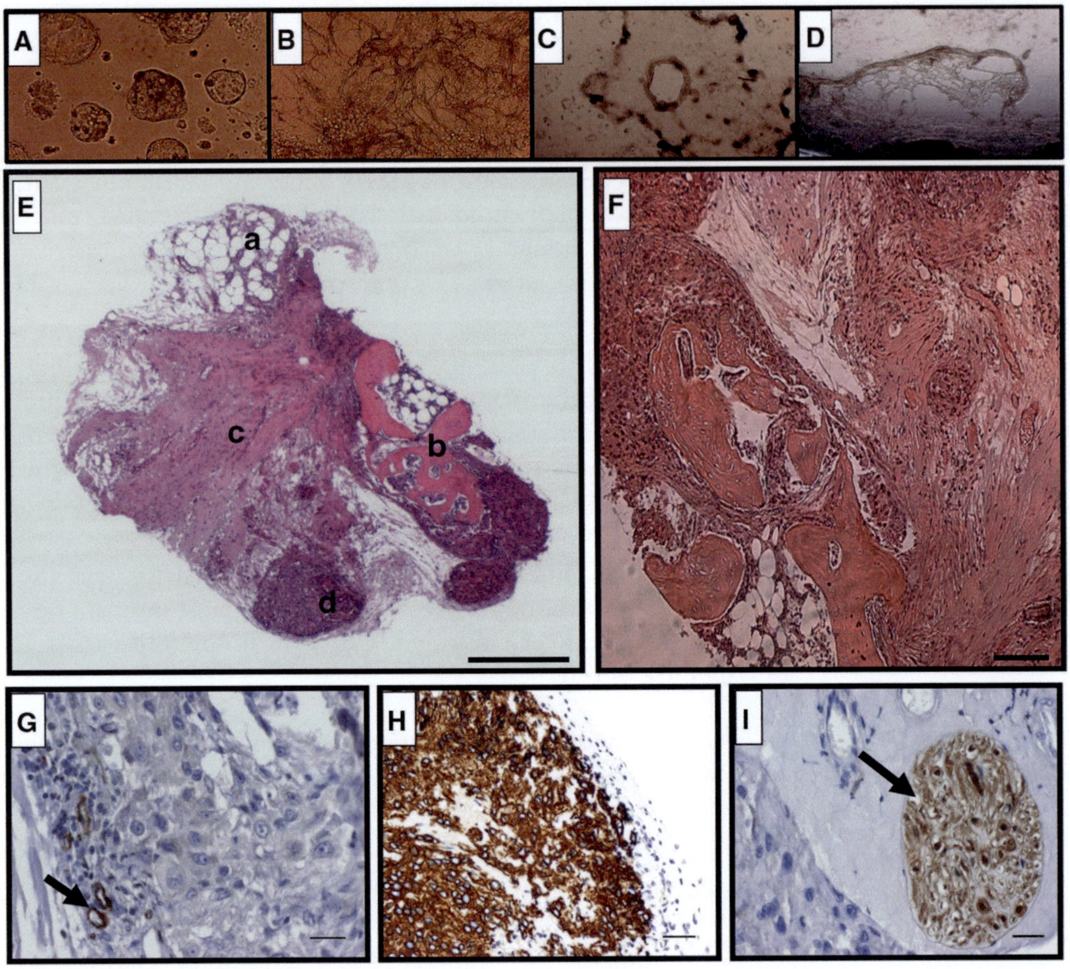

Fig. 5 Goat ES undergo spontaneous differentiation in vitro and can form complex teratomas in vivo. (**a**) Formation of goat ES cell-derived embryoid bodies from P25 gES cells. Derivation of neuronal (**b**) and epithelial (**c**) cells from goat ES-like cell differentiation. (**d**) Dissociated structure from ES-like cells. (**e**) General structure of gES teratoma comprising (*a*) adipose tissues, (*b*) bone marrow, (*c*) connective tissue, and (*d*) epithelium. Bar = 500 μm. (**f**) Larger size of bone and bone marrow from gES teratoma. Bar = 200 μm. (**g**) Blood vessel walls in goat ES cell-derived teratoma expressed desmin (*arrow* indicates desmin). Indirect immunohistochemistry and hematoxylin staining. Bar = 60 μm. (**h**) Epithelial portion of goat ES cell-derived teratoma strongly expressed intracytoplasmic cytokeratins AE1 and AE3. Indirect immunohistochemistry and hematoxylin staining. Bar = 60 μm. (**i**) Goat ES-derived teratoma nerve expressing S100 protein (*arrow* indicates nerve expressing S100 protein). Indirect immunohistochemistry and hematoxylin staining. Bar = 60 μm (adapted from ref. 12)

5. Anesthetize 7- to 8-week-old non-obese diabetic severe combined immunodeficient (NOD-SCID) mice using isoflurane vaporized in oxygen at a constant flow of 1–2 %. Aseptically prepare surgical sites.
6. Create the appropriate incisions for each injection location. Draw cell mixture immediately before injection using a 28.5-gauge needle and 1 ml syringe; inject contents from each tube

into the corresponding location in each mouse. Close the incisions. Follow protocols as described [18].

(a) Subcutaneous: Inject 200 μl cell mixture into the dorsolateral area into the subcutaneous space on both sides.

(b) Intramuscular: Make skin incision on left and right thigh; inject 50 μl cell mixture into central part of *biceps femoris* muscle.

(c) Testicular capsule: Make a 1-cm-long, transverse skin incision across the posterior abdomen, approximately over the location of the urinary bladder. Push the testicles slightly into the abdominal cavity to enable observation through the semitransparent abdominal wall. Inject 100 μl cell mixture through the abdominal wall and into the central part of the testicle.

7. To enhance cell engraftment post injection, allow the cell mixture to solidify in the localized transplantation site, and keep the mice under anesthesia on a 37 °C heat pad for 10–20 min after injection before allowing the mice to wake up. Allow mice to recover on a 37 °C heat pad.
8. Check mice 2–3 times a week. Monitor palpable tumors.
9. After 6–10 weeks from injection, sacrifice mice and harvest palpable tumors from the mice. Fix tumors in 10 % buffered formalin or 4 % paraformaldehyde, dehydrate, and embed in paraffin.
10. Obtain 5 μm serial sections for histological (hematoxylin and eosin) and indirect immunohistochemical analysis (for endodermal, mesodermal, ectodermal markers) (*see* Fig. 5).
11. Follow protocol for immunohistochemistry [12]. Treat tissues with hydrogen peroxide for 10 min to quench endogenous peroxidase activity. Incubate with primary murine monoclonal antibodies at dilutions of 1:500 for AE1/AE3, 1:30 for Vimentin V9, 1:40 for Desmin 1:40, and 1:800 for S-100. Perform incubation with primary antibodies at room temperature for 30 min and rinse with PBS. Apply Biotinylated Link (Dako, Burlington, ON, Canada), incubate for 15 min, and rinse with PBS. Apply Streptavidin–HRP (Dako, Burlington, ON, Canada), incubate for 15 min, and rinse with PBS. Counterstain tissue sections with Mayer's hematoxylin for 1 min at room temperature, and rinse with water. Dehydrate tissue and mount with Permount mounting media. Examine under a light microscope.
12. Assess histological sections of the tumor for tissue structures representative of different germ layers, which can be confirmed by the detection of specific tissue such as neuron, bone and cartilage (ectoderm), epithelial cells (endoderm), and skeletal and smooth muscle cells (mesoderm).

4 Notes

1. Leaving the gES cells at 3–5 cell clusters will allow the cryoprotectants to enter the gES cells and allow better recovery after thawing. It is also suggested that ROCK inhibitor may improve the survival of embryonic stem cells [19].
2. Passage number is one of the functional requirements of ES cells. True ES cell lines should maintain normal karyotype at higher passage numbers.
3. The CEeGFP plasmid used was provided by Dr. T. Takada, National Children' Medical Research Center, Tokyo, Japan.

Acknowledgements

This work was supported by Discovery Grant 385837-2010 and a Discovery Accelerator Supplement from the National Science and Engineering Research Council (NSERC) of Canada.

References

1. Evans MJ, Kaufman MH (1981) Establishment in culture of pluripotential cells from mouse embryos. Nature 292:154–156
2. Thomson JA, Kalishman J, Golos TG et al (1995) Isolation of a primate embryonic stem cell line. Proc Natl Acad Sci U S A 92:7844–7848
3. Thomson JA, Itskovitz-Eldor J, Shapiro SS et al (1998) Embryonic stem cell lines derived from human blastocysts. Science 282:1145–1147
4. Stice SL, Strelchenko NS, Keefer CL et al (1996) Pluripotent bovine embryonic cell lines direct embryonic development following nuclear transfer. Biol Reprod 54:100–110
5. Mitalipova M, Beyhan Z, First NL (2001) Pluripotency of bovine embryonic cell line derived from precompacting embryos. Cloning 3:59–67
6. Wang L, Duan E, Sung LY et al (2005) Generation and characterization of pluripotent stem cells from cloned bovine embryos. Biol Reprod 73:149–155
7. Cibelli JB, Stice SL, Golueke PL et al (1998) Cloned transgenic calves produced from non-quiescent fetal fibroblasts. Science 280: 1256–1258
8. Van Eijk MJT, van Rooijen MA, Modina S et al (1999) Molecular cloning, genetic mapping, and developmental expression of bovine POU5F1. Biol Reprod 60:1093–1103
9. Kirchhof N, Carnwath JW, Lemme E et al (2000) Expression pattern of Oct-4 in preimplantation embryos of different species. Biol Reprod 63:1698–1705
10. He H, Pant D, Schiffmacher A et al (2006) Developmental expression of pluripotency determining factors in caprine embryos: novel pattern of NANOG protein localization in the nucleolus. Mol Reprod Dev 73:1512–1522
11. Meinecke-Tillmann S, Meinecke B (1996) Isolation of ES-like cell lines from ovine and caprine pre-implantation embryos. J Anim Breed Genet 113:413–426
12. He S, Plant D, Bischoff S, Gavin W, Melican D, Keefer CL (2005) Expression of pluripotency-determining factors Oct-4 and Nanog in pre-implantation goat embryos. Reprod Fert Dev 17:203
13. Behboodi E, Bondareva A, Begin I et al (2011) Establishment of goat embryonic stem cells from in vivo produced blastocyst-stage embryos. Mol Reprod Dev 78:202–211
14. Garg S, Dutta R, Malakar D et al (2012) Cardiomyocytes rhythmically beating generated from goat embryonic stem cell. Theriogenology 77:829–839
15. De AK, Malakar D, Akshey YS et al (2011) Isolation and characterization of embryonic stem cell-like cells from in vitro produced goat

(*Capra hircus*) embryos. Anim Biotechnol 22: 181–196

16. Baldassarre H, Karatzas CN (2004) Advanced assisted reproduction technologies (ART) in goats. Anim Reprod Sci 82–83:255–266
17. Melican D, Gavin D (2008) Non-surgical embryo retrieval utilizing repeat superovulation and surgical embryo transfer in transgenic dairy goats. Theriogenology 69: 197–203
18. Prokhorova TA, Harkness LM, Frandsen U et al (2009) Teratoma formation by human embryonic stem cells is site dependent and enhanced by the presence of matrigel. Stem Cell Dev 18:47–54
19. Martin-Ibanez R, Unger C, Stromberg A et al (2008) Novel cryopreservation method for dissociated human embryonic stem cells in the presence of a ROCK inhibitor. Hum Reprod 23:2744–2754

Chapter 6

Derivation and Culture of Canine Embryonic Stem Cells

Ian C. Tobias, Courtney R. Brooks, Jonathan H. Teichroeb, and Dean H. Betts

Abstract

The derivation of canine embryonic stem cells (cESCs) represents a significant achievement and opens the door to further stem cell research and therapies in the dog. Canines share a common environment with humans and exhibit a host of genetic diseases, many of which have human parallels. Thus, the canine model presents unique advantages over other currently used organisms to help develop stem cell therapies in humans. To reveal the therapeutic potential of cESCs further basic research on the molecular mechanisms controlling their pluripotency and self-renewal characteristics is needed. Herein, we present the methods for derivation and culture of cESCs. Following collection of the canine blastocyst, two derivation methods are presented; immunodissection and whole blastocyst explant. These two methods lead to cESCs differing in morphology and subculture techniques. Additional protocols for subculture of established lines, feeder-free culture, and cryopreservation protocols are also described.

Key words Blastocyst, Canine, cESC derivation and culture, Embryonic stem cells, Pluripotency

1 Introduction

Embryonic stem cells (ESCs) are pluripotent cells with the ability to differentiate into all adult cell types from the three germ layers [1, 2]. This capacity may be observed by the formation of embryoid bodies in vitro or by teratoma development following injection into immunodeficient mice [3]. In the appropriate microenvironment, ESCs have unlimited self-renewal and therefore proliferate indefinitely [1, 2]. With these properties, ESCs show great promise for medical applications in tissue regeneration and the treatment of degenerative diseases [4] and are critical to the study of cellular lineage commitment [5]. To overcome the ethical [6] and immunological [7] complications of ES cell therapy, somatic cells can be genetically reprogrammed to a pluripotent state by the exogenous expression of Oct4, Sox2, Klf4, and c-Myc. These induced

Ramiro Alberio (ed.), *Epiblast Stem Cells: Methods and Protocols*, Methods in Molecular Biology, vol. 1074,
DOI 10.1007/978-1-62703-628-3_6, © Springer Science+Business Media New York 2013

pluripotent stem cells (iPSCs) provide an alternative source of undifferentiated cells which can be autologously transplanted into the original donor [8, 9].

The first ESC lines were isolated from the inner cell mass (ICM) of mouse preimplantation blastocysts [1]. Since then ESCs have been derived from the human embryo [4] and other mammalian embryos such as canine [10–14], rabbit [15], bovine [16], and porcine [17]. Although much of the work pertaining to the medical applications of ESCs has been done in rodent models [18], human ESCs (hESCs) differ considerably and therefore the methods developed in rodents do not always translate. Murine ESCs (mESCs) are dependent on STAT3 signaling via the presence of leukemia inhibitory factor (LIF) for self-renewal, and on bone morphogenic protein 4 (BMP4) to maintain the undifferentiated state [19, 20]. However, hESCs rely on basic fibroblast growth factor (bFGF or FGF2) stimulation of activin/nodal pathway to sustain pluripotency in long-term culture [21]. Morphologically, mESCs are propagated as tightly packed dome-shaped colonies [3], in contrast to the flattened appearance of ESCs derived from human embryos [22]. The cellular replicative demand on ESCs derived from large mammals with a relatively long life span extrapolate well to that of hESCs in terms of cell kinetics [23]. As a result, canine embryonic stem cells (cESCs) are more appropriate models of hESC therapies than those of a mouse or a rat.

Canines, being a longer lived species than murine, may be more appropriate for studies involving slowly progressive diseases or gerontological effects. Extensive lineage records, homogenous genetics, and pedigree-specific diseases arising from extensive in-breeding make them attractive as model organisms [24]. Many inherited human diseases have been found to possess canine parallels [25] and the dog has become a robust preclinical model for a range of human diseases. More recently, the development of canine iPS cells [9, 26–28] has opened the door for individualized tissue progenitor grafting studies in the dog. While technologies developed from the study of canine biology will contribute to the implementation of hESC therapeutics, they will also have medical and commercial importance in canine veterinary science.

1.1 Naïve and Primed Pluripotent States

ESCs derived from pre-blastocyst-stage embryos are fundamentally akin to cell lines established from the blastocyst [29]. Interestingly, cells isolated from murine postimplantation epiblasts show the hallmarks of pluripotency including the expression of the core pluripotent markers Oct4, Sox2, and Nanog and the ability to generate lineages from all three germ layers in vitro and in vivo [22, 30]. These murine epiblast stem cells (mEpiSCs) have features which are distinct from their mESC counterparts including long-term culture requirements, morphology, the ability to contribute to chimeras, expression levels of ICM-related genes, susceptibility to extraembryonic lineage commitment, X chromosome epigenetic status in female cells, and metabolic state [3, 22, 30, 31]. Although nonidentical, mEpiSCs exhibit qualities which resemble hESCs [30] challenging the original suggestion that the differences observed between mouse and human ESCs are attributable to inherent differences between the species. Instead, it is proposed that both the stage of embryonic development [32] and the ability of a defined microenvironment to stabilize pluripotent cells [33] may account for the different characteristics exhibited by immature (naïve) and mature (primed) cells.

Naïve mESC-like cells can be modified to a more primed EpiSC- or hESC-like state by in vitro conditions which support hESCs, but are unsuitable for immature cells [33]. Moreover, primed cells can revert back to a naïve state by the constitutive induction of exogenous transcription factors and/or the addition of two chemical inhibitors (2i) against glycogen synthase kinase 3β (GSK3β) and the mitogen-activated protein kinase (ERK1/2) pathway [34–36]. By interfering with signaling pathways which promote differentiation, these compounds support the propagation of immature ESCs [37] and even aid in reprogramming somatic cells into iPSCs [38]. Interestingly, environments with a low-oxygen tension seem to promote the conversion of naïve to primed cells [39].

1.2 Heterogeneity of Derived Canine Embryonic Stem Cells

Wilcox et al. [13] and other groups [10–12, 14] have isolated pluripotent cells from the ICM of the canine blastocyst. Though there are some discrepancies with certain hallmarks, each cESC line reported by the authors either expressed or stained positively for markers indicative of pluripotency such as Oct4, Nanog, alkaline phosphatase activity, and others. Likewise, each of the studies was able to demonstrate in vitro lineage commitment and differentiation into a variety of cell types. However, to date, only the cESC lines established by Vaags et al. and Wilcox et al. were capable of forming perceived teratomas in vivo, upon transplantation into immunodeficient mice [12, 13].

Of particular interest, cESC colonies seem to exhibit two distinct morphologies [23], which may be at least partly dictated by the methods used to derive the cells [13]. Similar to cell lines derived by Hayes and colleagues the cESC colonies isolated by immune-dissection comprised two discrete cell types, a dense central button of ESC-like cells surrounded by cells of a more flattened morphology [13]. These pluripotent cells were successfully propagated by Wilcox et al. in medium supplemented with LIF and bFGF [13], in contrast to the cESCs derived by Hayes et al. which could not be cultured long term in medium containing LIF only [11]. The remainder of the groups were able to derive pluripotent cells morphologically resembling those of hESC colonies [10, 12, 14], which are consistent with the description of colonies isolated from whole canine blastocyst explants [13]. These hESC-like cESCs also differed between authors in environmental factor dependency, as propagation in long-term culture required supplementation with either LIF [13, 14] or LIF and bFGF [12, 40]. Further research in canine ESC derivation and culture including the generation of canine induced pluripotent stem cells (ciPSCs) will help elucidate the pluripotency and self-renewal signaling pathways of these unique cells.

2 Materials

2.1 Canine Embryonic Stem Cell Culture Medium

The materials/reagents required for the maintenance of canine embryonic stem cells are indicated in Table 1.

2.1.1 Solutions for Use in Cell Culture

1. 100× L-glutamine stock (200 mM) solution: Add 2.92 g of L-glutamine to 90 mL of sterile water. Mix well until the solid has dissolved, and then add sterile water to 100 mL.
2. Stock bFGF solution: Reconstitute 100 μg of human bFGF in 5 mL of sterile water and mix well until solid has dissolved (*see* **Note 1**).

Table 1
Materials and reagents required for the maintenance of canine embryonic stem cells

Material/reagent	Supplier	Cat. no.
Nonessential amino acids	Invitrogen	11140-050
KnockOut Dulbecco's modified eagle medium (DMEM)	Invitrogen	10829-018
KnockOut serum replacement	Invitrogen	10828-028
Stem cell-qualified fetal bovine serum (FBS)	Invitrogen	10439-024
Human recombinant leukocyte inhibitory factor (hrLIF)	Sigma Aldrich	L5283
Beta-mercaptoethanol (BME)	Invitrogen	21985
L-Glutamine	Sigma Aldrich	G8540
Basic fibroblast growth factor (bFGF)	Invitrogen	PHG0021
EmbryoMax Nucleosides	Millipore	ES-008-D
Penicillin–Streptomycin	Invitrogen	15140
250 mL Stericup filter unit	Millipore	SCVP402RE
CF-1 mouse embryonic fibroblasts (MEFs)	GlobalStem	GSC-6301G
Dulbecco's phosphate-buffered saline (DPBS)	Invitrogen	14190-144
TrypLE Express	Invitrogen	12604-013
Geltrex basement membrane	Invitrogen	A1413301
1.5 mL Cryo-vials	Sigma	V4381-50EA

Table 2
Reagents used for embryo immunodissection

Material/reagent	Supplier	Cat. no.
Pronase	Sigma	P8811
Anti-canine polyclonal rabbit antiserum	Rockland Inc.	104-4101
Calcium and magnesium free Hank's balanced salt solution	Sigma	H9394
4-Well culture dishes	Nunc	176740
6-Well culture dishes	Nunc	140685
Dissecting microscope	–	–
Centrifuge	–	–
5 μL Wiretrol micropipette	Fisher	21-175A
EDTA	–	–
Gelatin	Sigma	G9391
Rubber policeman	–	–
Synthetic oviductal fluid (SOF)	–	–
35 mm Bacteriological petri dish	BD Biosciences	351008

3. Serum-free cESC base media: 85 mL of KnockOut DMEM, 15 mL of 100 % KnockOut serum replacement, 1 mL of 200 mM L-glutamine, 0.5 mL of 100× nonessential amino acids, 0.5 mL of 100× (10,000 U/mL) penicillin–streptomycin, 0.5 mL of 100× nucleoside preparation (*see* **Note 2**), 100 μL of 55 mM beta-mercaptoethanol (*see* **Note 3**).
4. Serum-free cESC culture media (ES+): Add 100 μL of human recombinant (hr) LIF 10 μg/mL stock solution and 20 μL of FGF 20 μg/mL stock to 100 mL of filtered cESC base media (*see* **Note 4**).
5. 10 % FBS in cESC culture media (ES++): Dilute ESC-qualified FBS 1:10 (v/v) in cESC culture media.

2.2 Immuno-dissection

The reagents used for embryo immunodissection are indicated in Table 2.

2.2.1 Solutions for Use in Immunodissection

1. 0.5 % Pronase solution: Reconstitute 250 mg of pronase in 50 mL of sterile water.
2. 0.1 % Gelatin solution: Add 1 g of gelatin to 1 L of deionized water, and then autoclave the solution to dissolve the solid.
3. 100× EDTA stock (3 mg/mL) solution: Add 0.3 g of EDTA to 95 mL of deionized water, stirring well to dissolve. Add sterile water to 100 mL.

Table 3
Materials used for embryo explants

Material/reagent	Supplier	Cat. no.
Collagenase Type IV	Sigma	C5138
35 mm bacteriological petri dish	BD Biosciences	351008
5 μL wiretrol micropipette	Fisher	21-175A
Dissecting microscope	–	–
6-Well culture dishes	Nunc	140685
4-Well culture dishes	Nunc	176740
Gelatin	Sigma	G9391

2.3 Explant Methodology

The materials used in the embryo explants method are presented in Table 3.

2.3.1 Solutions for Use in Embryo Explant Methods

1. 0.1 % (200 U/mL) Collagenase IV: Reconstitute 50 mg of collagenase IV in 50 mL of sterile water.

3 Methods

3.1 Establishment of Canine Embryonic Stem Cell Colonies from Immunodissection Inner Cell Masses

In brief, canine blastocysts are procured from mature bitches, specifically hounds and beagles. Mature bitches are observed for signs of heat such as intravaginal edema and mounting before inseminating bitches in estrus once or twice. Progesterone and luteinizing hormone (LH) levels are closely monitored for an LH surge or a progesterone concentration greater than 16 nmol/L to define post-ovulation day 0. At days 11–15 post LH surge, ovario-hysterectomy should be performed and each uterine horn flushed downward near the uterotubal junction with 5–10 mL of ViGRO Complete Flush Solution using a syringe. Canine embryos are collected using a sterile 10 cm culture dish under a stereoscope. Uterine horn flushing may be repeated before retrieving blastocysts. Canine morulae or blastocysts may be treated directly or cultured overnight in synthetic oviductal fluid (SOF) at an incubation temperature of 38.5 °C under 5 % CO_2 and 5 % O_2 to allow for further development or re-expansion, respectively (*see* **Note 5**).

Embryological immunosurgery is typically a two-step process which relies on the impermeability of the ICM to particular antiserum preparations. The subsequent addition of complement causes lysation of all antibody-bound cells. Selective trophoblastic cell cytotoxicity allows for swift mechanical removal of the ICM [41]. However, it is possible to digest the trophoblast and isolate the

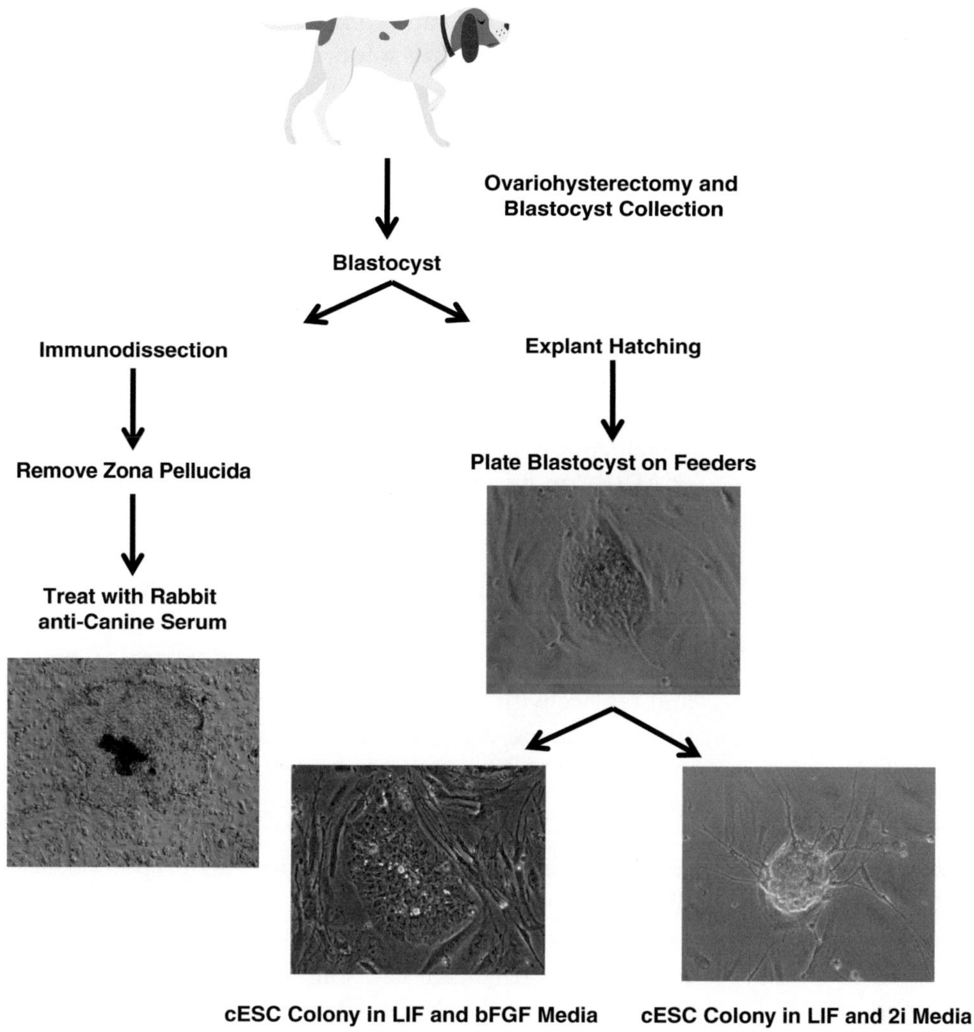

Fig. 1 Schematic depiction of the steps involved in the derivation and culture of canine embryonic stem cells. Canine blastocysts procured are subject to immunodissection or embryo explant protocols and exhibit two distinct morphologies. Immunodissected colonies possess a dense central "button" surrounded by flattened cells. Explant colonies appear hESC-like, but morphological changes associated with a more naïve pluripotent-like state are observed upon propagation of explanted colonies in media containing LIF or LIF + 2i (glycogen synthase kinase 3β and mitogen-activated protein kinase inhibitors)

ICM from the embryo using the antiserum alone (*see* Fig. 1 for schematic overview of methods).

1. If not already hatched, remove the zona pellucida from the live morula or blastocyst-stage embryos using 0.5 % pronase solution in a non-adherent 35 mm bacteriological petri dish for approximately 2 min (*see* **Note 6**). Remove the embryo from the pronase solution by pipette and place in a sterile non-adherent dish. Wash two to three times with SOF.

2. Dilute the polyclonal antiserum 1:10 in calcium- and magnesium-free Hank's balanced salt solution (*see* **Note 7**). Under a dissecting microscope, expose the embryo to the anti-canine antibody preparation for 1–5 min at room temperature.
3. During the treatment, use a small pulled glass pipette to agitate the trophoectoderm with the antibody preparation. Then dissociate the loosened trophoectoderm using a sharp wiretrol micropipette. Wash two to three times with SOF.
4. Slice the ICM with the wiretrol into small loose clumps of cells (*see* **Note 8**).
5. Plate the clumps of presumptive ICM on monolayers of arrested feeder cells seeded at 10^5 cells/cm^2 on gelatin-coated 4-well culture dishes. For the culture media use a 1:1 mixture of MEF-conditioned cESC culture media and fresh cESC culture media supplemented with stem cell-qualified serum (*see* **Note 9**).
6. Incubate the cells overnight at 38.5 °C with 5 % CO_2, being careful not to disturb the plate while the cells are adhering.
7. Following overnight incubation, aspirate the attachment media and replace with serum-free cESC culture media. Perform half media changes every 2 days and passage the cells every 4–7 days (when at 20 % confluency) until purified.

3.2 Initial Subculture of Immunodissected Canine Embryonic Stem Cells

1. Prepare growth-arrested MEF plates by seeding γ-irradiated MEFs at a confluency of 10^5 cells/cm^2 into an appropriate culture dish one day ahead of subculture.
2. Dilute EDTA stock solution 1:100 in calcium- and magnesium-free Hank's balanced salt solution. Add 1 mL of working EDTA solution (*see* **Note 10**) to the culture dishes and incubate for 5–10 min at room temperature.
3. If there are only a few presumptive cESC colonies, remove the individual colonies mechanically using a wiretrol micropipette and place them in a new 4-well culture dish with EDTA for a few more minutes of digestion. If there are many colonies (and low levels of differentiation) scrape the entire plate with a rubber policeman, and then break up the colonies with gentle pipetting of the suspension.
4. Wash the colonies by gentle centrifugation in a 15 mL conical tube at 150 RCF and resuspend the pellet in a 1:1 mixture of serum-free MEF-conditioned cESC culture media and culture media supplemented with 10 % qualified FBS.
5. Reseed the cell suspension in a 4-well plate at split ratios of 1:1 or 1:2 for dishes with a low colony density or at split ratios between 1:3 and 1:7 for dishes with a high colony density on growth-arrested mouse embryonic fibroblasts.

3.3 Isolation of Canine Embryonic Stem Cell Colonies from Outgrowths of Intact Embryo Explants

1. Plate morula or blastocyst embryos on growth-arrested feeder cells seeded at 10^5 cells/cm^2 on gelatin-coated 4-well dishes. Incubate the embryos at 38.5 °C in a 5 % CO_2 environment, checking the explants daily for outgrowth.
2. Maintain the embryos on the feeder layer for 1–3 days using cESC media until hatched or fully expanded. If embryos have not hatched by 3 days, nick the zona pellucida with a sharp pulled glass pipette to allow the ICM to grow out.
3. Observe the growth of the explants daily, then excise the ICM from the monolayer using a wiretrol micropipette, and plate into a fresh culture dish.
4. Perform half cESC culture media changes every 2 days and subculture the presumptive ICM every 4–7 days when at about 20 % confluent until purified.

3.4 Initial Subculture of Explanted Canine Embryonic Stem Cells

1. For culture dishes with only a few colonies, selectively cut out the undifferentiated colonies from the feeder layer under a dissecting microscope with a wiretrol micropipette. For culture dishes with many colonies and low levels of differentiation, add 0.7 mL of 0.1 % collagenase Type IV per well of a 6-well dish (or 0.35 mL per well of a 4-well dish) and incubate cESCs for 20–30 min (*see* **Note 11**). Scoop out the colonies under a dissecting microscope.
2. Transfer the colonies into a 35 mm non-adherent petri dish containing fresh 10 % ES-qualified FBS in cESC culture media.
3. Mechanically cut the colonies into 2–8 pieces, depending on the size of the original colony using a wiretrol micropipette.
4. Recover colony fragments and transfer them onto new growth-arrested feeder layers in a multiwell dish. Add 30–50 fragments per 35-mm well and 2–5 fragments per 19-mm well.
5. Rinse the non-adherent cutting dish with cESC culture media to recover any residual colony fragments and transfer to the new feeder dishes as well.
6. Check the level of colony fragment adherence the following day as some of the larger fragments may have difficulty attaching (*see* **Note 12**).

3.5 Long-Term Culture of Canine Embryonic Stem Cells

1. Perform half cESC culture media changes daily with full media changes every 3 days and passage onto a new feeder layer about every 7–8 days, or when colonies are about to touch.
2. Prepare growth-arrested MEF plates by seeding γ-irradiated MEFs at a confluency of 10^5 cells/cm^2 into an appropriate culture dish 1 day prior to cell passage.
3. Aspirate cESC culture media and wash cells once in DPBS without calcium and magnesium. Add 0.5–1 mL of TrypLE Express and gently shake the culture dish to help detach cESC

colonies from the feeder layer. Monitor for rounding of cells and detachment of the edges of colonies.

4. Once colony edges begin to lift, dissociate colonies by rinsing with MEF media. Very gently pipette the cells several times to break up colonies and centrifuge at 150 RCF in a 15 mL conical tube with a volume of MEF media at least 2× the volume of TrypLE Express to inactivate the enzymes.
5. Aspirate the supernatant and resuspend cESC pellet in culture media (*see* **Note 13**) and gently pipette the solution in the tube a few times to break up colonies into smaller fragments.
6. Seed the cell suspension into new MEF layers at a ratio of between 1:2 and 1:4. Do not disturb the colonies while adhering overnight in an incubator at 38.5 °C and 5 % CO_2.

3.6 Feeder-Free Culture of Canine Embryonic Stem Cells

1. Prepare growth-arrested MEF plates for media conditioning by seeding γ-irradiated MEFs at a confluency of 10^5 cells/cm^2 into a 15 cm culture dish. 12–24 hrs after plating aspirate the MEF media and add 30 mL of base cESC culture media (without growth factors).
2. After 48 h, pipette the conditioned base cESC culture media into a flask for collection. Add hrLIF to 10 ng/μL and bFGF to 4 ng/μL, and filter through a Stericup before use.
3. Coat the growth surface of each culture dish well with Geltrex diluted 1:100 in KO DMEM for 60 min at 37 °C. Aspirate the media and seed cESC colonies at an appropriate confluency in MEF-conditioned cESC culture media.
4. Perform daily half MEF-conditioned media changes with a full media change every 3 days.

3.7 Cryopreservation

For long-term storage cESCs should be frozen at an early passage number. Standard cryopreservation techniques can be used as follows:

1. Prepare fresh 10 % DMSO/90 % FBS cryopreservation solution and cool it on ice.
2. Aspirate the media from one well of a 12- or a 6-well plate and rinse briefly with DPBS. Apply 0.5 mL to 1 mL TrypLE Express to well and monitor for rounding of cells and detachment of the edge of colonies (*see* **Note 14**).
3. Once colony edges begin to lift, use a stream of pipetted MEF media to dislodge the colonies. Very gently pipette the cells several times to break up colonies and centrifuge at 150 × *g* in a 15 mL tube with a volume of MEF media at least 2× the volume of TrypLE Express to inactivate the enzymes.
4. Aspirate the supernatant and gently resuspend pellet in approximately 1 mL of cold cryopreservation solution, depositing into a pre-chilled cryo-tube.

Table 4
Antibodies used to characterize canine embryonic stem cells

Pluripotency markers		Differentiation markers	
Marker	**Supplier (cat. no.)**	**Marker**	**Supplier (cat. no.)**
Oct4	Santa Cruz (SC9081)	Gata6	R&D (AF1700)
Sox2	Stem Cell Tech (1438)	TUBB3/TUJ1	Sigma (T8660)
Nanog	Peprotech (500-P236)	GFAP	Millipore (MAB360)
Alkaline Phosphatase	Vector Lab (SK5100)	NCAM	Millipore (AB5032)
SSEA-1	Millipore (MAB4301)	Tromal	DSHB
SSEA-3	Millipore (MAB4303)	AFP	Dako(A0008)
SSEA-4	Millipore (MAB4304)	MAP2	Millipore (AB5622)
TRA-1-60	Millipore (MAB4360)	vWF	Millipore (AB7356)
TRA-1-81	Millipore (MAB4381)	CXCR4	Gene Tex (GTX10403)
Rexl	Abcam (ab50828-50)	FLK-1/VEGFR2	Millipore (07–716)
H3K27me3	Abcam (ab6002)	IB4	Invitrogen (I21411)
Lin28	Santa Cruz (67266)	β-III-tubulin	Millipore (CBL412)
Klf4	Abcam	CD201	BD Biosciences (557950)
c-MYC	Abcam	CCSP	Millipore (07–623)
		Nestin	DSHB (RAT-401)
		Vimentin	Millipore (MAB3400)
		Sox17	R&D (AB2132)
		Desmin	Sigma

5. Freeze cells at approximately 1 °C/min in an isopropanol cell freezing container.
6. After 24 h transfer cells to liquid nitrogen storage.

3.8 Characterization of Canine Embryonic and Induced Pluripotent Stem Cells

The cESCs and canine iPSCs described in the literature to date are immunopositive with canine-reactive antibodies for the following markers of pluripotency and differentiation (Table 4).

3.9 Future Perspectives and Conclusions

Although numerous groups have successfully derived cESC, the protocols used to establish these lines vary. The differences in cESC derivation and culture likely contribute to the heterogeneous descriptions provided by the authors [10–14]. This is likely due to a lack of understanding of pluripotent cell development in canine reproductive physiology, and the pathways controlling cESC pluripotency and self-renewal. It is important that these pathways are elucidated and that the methods of cESC isolation and culture are optimized. The fact that cESCs, despite demonstrating in vitro differentiation into multiple lineages, have performed poorly in teratoma assays warrants further investigation [23]. Although body temperature differences between species may be the underlying root to this problem, suspension of cESCs or canine iPSCs in a mixture of KnockOut DMEM, Matrigel, and collagen injected intramuscularly into a more immunocompromised mouse strain (NOD/SCID IL-2Rγ nulls) might prove successful [42]. The ultimate test of competency in mESCs and murine iPSCs has been that of the ability to contribute to a chimeric embryo [3]. Production of a proper and viable embryo during tetraploid complementation is the most discerning test to date of the naïvety and pluripotency of ESCs and iPSCs. Whether current generations of cESCs can contribute to an embryo remains to be determined.

Although the transition between naïve and primed pluripotent states has been investigated in both mouse and human ESCs [34, 43], modulation of cESC pluripotency is in its early stages. Our preliminary studies have shown that cESCs are responsive to low-oxygen conditions in LIF + 2*i*-supplemented media with colonies displaying domed-like morphology typical of naïve pluripotency (*see* Fig. 1). Developing this biotechnology in the canine model has obvious veterinary importance and since many inherited canine pathologies have a human equivalent [25], studies delineating the regulation of cESC pluripotency are essential to the progression of regenerative medicine. It is possible that low-oxygen microenvironments may facilitate cESC self-renewal. Hypoxia-inducible factors (HIFs) regulate pluripotency and proliferation in human ESCs cultured at reduced oxygen tensions [44]. It is also likely that 38.5 °C, while permissive to the cESCs, places the MEF layer in a constant state of heat shock. This may affect feeder layer secretion of essential growth factors [45] and therefore the microenvironmental niche in which cESCs are propagated. This issue may be resolved by maintaining cESCs under feeder-free conditions. Thus, improvements in the derivation and culture of cESCs will not only decipher the molecular mechanisms of pluripotency and self-renewal but will also lead to the development of stem cell therapies/technologies for both human and canine patients.

4 Notes

1. It is recommended that the bFGF stock solution is stored at −20 °C in working aliquots to avoid repeated freeze–thaw cycles that degrade the protein.

2. The liquid preparation of nucleosides contains 3 mM of adenosine, cytosine, guanidine, uridine, and 1 mM of thymine.
3. Prepare the base cESC media every 3 weeks to keep the ingredients fresh and ensure that all components are filtered through a Stericup unit before use.
4. Prepare culture media weekly and either avoid filtration of LIF and bFGF components or use low-protein-binding filters, adding KO serum first.
5. Isolated canine embryos and presumptive cESCs were cultured at 38.5 °C (the internal body temperature of the dog) rather than standard conditions as it was more permissive to cell growth.
6. Acid Tyrode's solution was unable to dissolve canine embryo zona pellucida.
7. The undiluted antiserum preparation may be stored in small frozen aliquots and thawed prior to use.
8. The ICM may also be passively broken up by gentle agitation in calcium- and magnesium-free Hank's balanced salt solution.
9. The use of serum aims to facilitate colony attachment only. Minimize serum exposure to prevent differentiation of the cells.
10. In general, the use of trypsin or collagenase for immunodissected cESC colonies is contraindicated; however, following an adaptation period of several passages, we have found that passaging by collagenase or TrypLE express is possible.
11. Do not expose the colonies to collagenase for too long as the colonies should remain intact. Do not attempt to generate a single-cell suspension.
12. Larger colony fragments may not adhere overnight and may need to be cut into the MEF layer to facilitate attachment the next day. To perform this, rotate the sharp edge of the wiretrol micropipette or pulled glass pipette over the colony in situ.
13. Once cell lines have been established cESC culture media no longer requires FBS supplementation to facilitate attachment, provided colonies are left undisturbed for 24 h to adhere.
14. While manual dissection of colonies can be performed for cryopreservation, the smaller colonies created during enzymatic passaging freeze more evenly and exhibit enhanced viability upon thawing.

Acknowledgments

The financial support of the Natural Sciences and Engineering Research Council of Canada (NSERC) and Canada Foundation for Innovation (CFI) are gratefully acknowledged.

References

1. Evans MJ, Kaufman MH (1981) Establishment in culture of pluripotential cells from mouse embryos. Nature 292:154–156
2. Martin GR (1981) Isolation of a pluripotent cell line from early mouse embryos cultured in medium conditioned by teratocarcinoma stem cells. Proc Natl Acad Sci U S A 78:7634–7638
3. Kuijk EW, de Sousa C, Lopes SM, Geijsen N, Macklon N, Roelen BA (2011) The different shades of mammalian pluripotent stem cells. Hum Reprod Update 17:254–271
4. Thomson JA, Itskovitz-Eldor J, Shapiro SS, Waknitz MA, Swiergiel JJ, Marshall VS, Jones JM (1998) Embryonic stem cell lines derived from human blastocysts. Science 282:1145–1147
5. Parsons XH, Teng YD, Parsons JF, Snyder EY, Smotrich DB, Moore DA (2011) Efficient derivation of human neuronal progenitors and neurons from pluripotent human embryonic stem cells with small molecule induction. J Vis Exp 56:e3273
6. Cregan K (2005) Ethical and social issues of embryonic stem cell technology. Intern Med J 35:126–127
7. Drukker M, Katz G, Urbach A, Schuldiner M, Markel G, Itskovitz-Eldor J, Reubinoff B, Mandelboim O, Benvenisty N (2002) Characterization of the expression of MHC proteins in human embryonic stem cells. Proc Natl Acad Sci U S A 99:9864–9869
8. Takahashi K, Yamanaka S (2006) Induction of pluripotent stem cells from mouse embryonic and adult fibroblast cultures by defined factors. Cell 126:663–676
9. Lee AS, Xu D, Plews JR, Nguyen PK, Nag D, Lyons JK, Han L, Hu S, Lan F, Liu J, Huang M, Narsinh KH, Long CT, de Almeida PE, Levi B, Kooreman N, Bangs C, Pacharinsak C, Ikeno F, Yeung AC, Gambhir SS, Robbins RC, Longaker MT, Wu JC (2011) Preclinical derivation and imaging of autologously transplanted canine induced pluripotent stem cells. J Biol Chem 286:32697–32704
10. Hatoya S, Torii R, Kondo Y, Okuno T, Kobayashi K, Wijewardana V, Kawate N, Tamada H, Sawada T, Kumagai D, Sugiura K, Inaba T (2006) Isolation and characterization of embryonic stem-like cells from canine blastocysts. Mol Reprod Dev 73:298–305
11. Hayes B, Fagerlie SR, Ramakrishnan A, Baran S, Harkey M, Graf L, Bar M, Bendoraite A, Tewari M, Torok-Storb B (2008) Derivation, characterization, and in vitro differentiation of canine embryonic stem cells. Stem Cells 26:465–473
12. Vaags AK, Rosic-Kablar S, Gartley CJ, Zheng YZ, Chesney A, Villagomez DA, Kruth SA, Hough MR (2009) Derivation and characterization of canine embryonic stem cell lines with in vitro and in vivo differentiation potential. Stem Cells 27:329–340
13. Wilcox JT, Semple E, Gartley C, Brisson BA, Perrault SD, Villagomez DA, Tayade C, Becker S, Lanza R, Betts DH (2009) Characterization of canine embryonic stem cell lines derived from different niche microenvironments. Stem Cells Dev 18:1167–1178
14. Schneider MR, Adler H, Braun J, Kienzle B, Wolf E, Kolb HJ (2007) Canine embryo-derived stem cells–toward clinically relevant animal models for evaluating efficacy and safety of cell therapies. Stem Cells 25:1850–1851
15. Wang S, Tang X, Niu Y, Chen H, Li B, Li T, Zhang X, Hu Z, Zhou Q, Ji W (2007) Generation and characterization of rabbit embryonic stem cells. Stem Cells 25:481–489
16. Mitalipova M, Beyhan Z, First NL (2001) Pluripotency of bovine embryonic cell line derived from precompacting embryos. Cloning 3:59–67
17. Li M, Zhang D, Hou Y, Jiao L, Zheng X, Wang WH (2003) Isolation and culture of embryonic stem cells from porcine blastocysts. Mol Reprod Dev 65:429–434
18. Banerjee ER, Laflamme MA, Papayannopoulou T, Kahn M, Murry CE, Henderson WR Jr (2012) Human embryonic stem cells differentiated to lung lineage-specific cells ameliorate pulmonary fibrosis in a xenograft transplant mouse model. PLoS One 7:e33165
19. Chambers I, Smith A (2004) Self-renewal of teratocarcinoma and embryonic stem cells. Oncogene 23:7150–7160
20. Ying QL, Nichols J, Chambers I, Smith A (2003) BMP induction of Id proteins suppresses differentiation and sustains embryonic stem cell self-renewal in collaboration with STAT3. Cell 115:281–292
21. Vallier L, Alexander M, Pedersen RA (2005) Activin/Nodal and FGF pathways cooperate to maintain pluripotency of human embryonic stem cells. J Cell Sci 118:4495–4509. doi:10.1242/jcs.02553
22. Brons IG, Smithers LE, Trotter MW, Rugg-Gunn P, Sun B, de Sousa C, Lopes SM, Howlett SK, Clarkson A, Ahrlund-Richter L, Pedersen RA, Vallier L (2007) Derivation of pluripotent epiblast stem cells from mammalian embryos. Nature 448:191–195

23. Schneider MR, Wolf E, Braun J, Kolb HJ, Adler H (2010) Canine embryonic stem cells: state of the art. Theriogenology 74:492–497
24. Shearin AL, Ostrander EA (2010) Leading the way: canine models of genomics and disease. Dis Model Mech 3:27–34
25. Starkey MP, Scase TJ, Mellersh CS, Murphy S (2005) Dogs really are man's best friend–canine genomics has applications in veterinary and human medicine! Brief Funct Genomic Proteomic 4:112–128
26. Luo J, Suhr ST, Chang EA, Wang K, Ross PJ, Nelson LL, Venta PJ, Knott JG, Cibelli JB (2011) Generation of leukemia inhibitory factor and basic fibroblast growth factor-dependent induced pluripotent stem cells from canine adult somatic cells. Stem Cells Dev 20:1669–1678
27. Shimada H, Nakada A, Hashimoto Y, Shigeno K, Shionoya Y, Nakamura T (2010) Generation of canine induced pluripotent stem cells by retroviral transduction and chemical inhibitors. Mol Reprod Dev 77:2
28. Whitworth DJ, Ovchinnikov DA, Wolvetang EJ (2012) Generation and characterization of LIF-dependent canine induced pluripotent stem cells from adult dermal fibroblasts. Stem Cells Dev 21:2288–2297
29. Tesar PJ (2005) Derivation of germ-line-competent embryonic stem cell lines from preblastocyst mouse embryos. Proc Natl Acad Sci U S A 102:8239–8244
30. Tesar PJ, Chenoweth JG, Brook FA, Davies TJ, Evans EP, Mack DL, Gardner RL, McKay RD (2007) New cell lines from mouse epiblast share defining features with human embryonic stem cells. Nature 448:196–199
31. Zhou W, Choi M, Margineantu D, Margaretha L, Hesson J, Cavanaugh C, Blau CA, Horwitz MS, Hockenbery D, Ware C, Ruohola-Baker H (2012) HIF1alpha induced switch from bivalent to exclusively glycolytic metabolism during ESC-to-EpiSC/hESC transition. EMBO J 31: 2103–2116
32. Nichols J, Smith A (2009) Naive and primed pluripotent states. Cell Stem Cell 4:487–492. doi:10.1016/j.stem.2009.05.015
33. Hanna J, Markoulaki S, Mitalipova M, Cheng AW, Cassady JP, Staerk J, Carey BW, Lengner CJ, Foreman R, Love J, Gao Q, Kim J, Jaenisch R (2009) Metastable pluripotent states in NOD-mouse-derived ESCs. Cell Stem Cell 4:513–524
34. Bao S, Tang F, Li X, Hayashi K, Gillich A, Lao K, Surani MA (2009) Epigenetic reversion of post-implantation epiblast to pluripotent embryonic stem cells. Nature 461:1292–1295
35. Guo G, Yang J, Nichols J, Hall JS, Eyres I, Mansfield W, Smith A (2009) Klf4 reverts developmentally programmed restriction of ground state pluripotency. Development 136:1063–1069
36. Hanna J, Cheng AW, Saha K, Kim J, Lengner CJ, Soldner F, Cassady JP, Muffat J, Carey BW, Jaenisch R (2010) Human embryonic stem cells with biological and epigenetic characteristics similar to those of mouse ESCs. Proc Natl Acad Sci U S A 107:9222–9227
37. Ying QL, Wray J, Nichols J, Batlle-Morera L, Doble B, Woodgett J, Cohen P, Smith A (2008) The ground state of embryonic stem cell self-renewal. Nature 453:519–523
38. Silva J, Barrandon O, Nichols J, Kawaguchi J, Theunissen TW, Smith A (2008) Promotion of reprogramming to ground state pluripotency by signal inhibition. PLoS Biol 6:e253
39. Takehara T, Teramura T, Onodera Y, Hamanishi C, Fukuda K (2012) Reduced oxygen concentration enhances conversion of embryonic stem cells to epiblast stem cells. Stem Cells Dev 21:1239–1249
40. Wilcox JT, Lai JK, Semple E, Brisson BA, Gartley C, Armstrong JN, Betts DH (2011) Synaptically-competent neurons derived from canine embryonic stem cells by lineage selection with EGF and Noggin. PLoS One 6:e19768
41. Solter D, Knowles BB (1975) Immunosurgery of mouse blastocyst. Proc Natl Acad Sci U S A 72:5099–5102
42. Hotta A, Cheung AY, Farra N, Vijayaragavan K, Seguin CA, Draper JS, Pasceri P, Maksakova IA, Mager DL, Rossant J, Bhatia M, Ellis J (2009) Isolation of human iPS cells using EOS lentiviral vectors to select for pluripotency. Nat Methods 6:370–376
43. Gu Q, Hao J, Zhao XY, Li W, Liu L, Wang L, Liu ZH, Zhou Q (2012) Rapid conversion of human ESCs into mouse ESC-like pluripotent state by optimizing culture conditions. Protein Cell 3:71–79
44. Szablowska-Gadomska I, Zayat V, Buzanska L (2011) Influence of low oxygen tensions on expression of pluripotency genes in stem cells. Acta Neurobiol Exp 71:86–93
45. Villa-Diaz LG, Ross AM, Lahann J, Krebsbach PH (2013) The evolution of human pluripotent stem cell culture: from feeder cells to synthetic coatings. Stem Cells 31(1):1–7. doi:10.1002/stem.1260

Chapter 7

Isolation and Culture of Porcine Embryonic Stem Cells

Ivan Vassiliev and Mark B. Nottle

Abstract

Despite their agricultural and biomedical importance, embryonic stem cells (ESCs) are yet to be isolated for the pig or the domestic ungulates in general. This suggests that methods which have been used successfully in mice may not be applicable to these. In this chapter we describe a new method for the isolation of porcine ESCs. This method differs from those described previously in that it produces homogeneous outgrowths from undifferentiated inner cell mass cells when embryos are plated onto inactivated mouse embryonic feeder layers.

Key words Embryonic stem cells, Inner cell mass, Epiblast, Pig

1 Introduction

Mammalian embryonic stem cells (ESCs) are pluripotent cells derived from pregastrulation embryos. These cells retain their capacity for self-renewal and can differentiate into all cell types found in the body, as demonstrated by teratoma formation and the production of chimeric animals following blastocyst injection and embryo aggregation. The production of chimeric animals together with the demonstration that these cells can give rise to gametes is considered to be the ultimate proof of pluripotency, but has only been demonstrated in mice and rats. ESCs have also been isolated in humans and nonhuman primates, but chimera formation has not been demonstrated in the latter [1]. Interestingly human and nonhuman primate ESCs may be more like mouse epiblast cells than naive mouse ESCs [2]. Although these cells are termed "pluripotent" they have limited developmental potential compared with ESCs. In particular they are highly inefficient in generating chimeras [2]. This may also explain why putative ESCs have been isolated for a variety of species but have failed to produce chimeras.

Attempts to isolate ESCs in pigs have so far proved elusive. Most workers have used methods similar to those employed initially in the mouse [3, 4], namely, the mechanical or the immunosurgical

Ramiro Alberio (ed.), *Epiblast Stem Cells: Methods and Protocols*, Methods in Molecular Biology, vol. 1074, DOI 10.1007/978-1-62703-628-3_7,

isolation of the inner cell mass (ICM) which is then plated onto an inactivated mouse embryonic feeder layers in the presence of serum. In mice, these outgrowths are trypsiniZed and replated in an attempt to generate colonies from the relatively few ESC progenitor cells present initially. Putative ESCs are selected normally on a morphological basis, and passaged to produce primary cell lines which are then frozen and thawed to determine their viability. However, this approach has been unsuccessful for the pig and domestic ungulates [5, 6] in general, suggesting that a different approach may be required.

In the pig the ICM of early blastocysts consists of relatively few cells which are morphologically indiscernible, preventing mechanical and immunosurgical isolation. As such, most studies have used fully expanded blastocysts or later stages where ICM has already begun to differentiate into the epiblast and hypoblast, resulting in heterogeneous outgrowths where these, as well as other differentiated cells, rapidly overgrow the few remaining ESC progenitor cells. To overcome this problem we deliberately chose to use blastocysts where the ICM is yet to differentiate. This necessitated using whole embryos because the ICM is not easily isolated. Because these embryos do not attach readily to the feeder layer at this stage, we chose to flatten or embed these onto the feeder cells to promote direct contact between the ICM cells and the feeder layer.

In developing our approach, our aim was to produce homogeneous outgrowths which do not differentiate. We reexamined a number of components used to isolate ESCs; in particular, we compared α-MEM with the more commonly used DMEM because of promising results obtained previously in cattle by us [7]. We also compared FCS with serum replacement, because the former is thought to promote differentiation. These experiments were done under low oxygen because in preliminary experiments we found that atmospheric oxygen also promoted differentiation. The results of this study showed that when blastocysts with an undifferentiated ICM are selected, the use of DMEM and FCS alone and combination with each other resulted in the production of heterogeneous outgrowths. In contrast, the combination of MEM and serum replacement resulted in homogeneous outgrowths at relatively high efficiencies [8]. The lack of serum attachment factors in serum replacement also resulted in more metabolically active trophectoderm cells detaching from the feeder layer and degenerating, leaving homogeneous pluripotent outgrowths. Mechanical dissociation into clumps was used to overcome any possible effects of trypsinization disrupting cell-to-cell contacts that promote differentiation [9]. In summary, by choosing early blastocysts where the ICM is yet to differentiate and replacing those components thought to promote differentiation, we have, for the first time to our knowledge, been able to produce homogeneous outgrowths of pig ESCs.

This novel feature has facilitated the isolation of putative pig ESC lines from in vitro-produced (IVP) and in vivo-derived porcine embryos [8, 10], as well as from somatic cell nuclear transfer or cloned embryos [11], and parthenogenetic embryos (unpublished results). These cells have been extensively characterized in vitro and shown to survive repeated passaging and cryopreservation without changes in pluripotent marker expression and karyotype. These cells can be directed to differentiate in vitro to cell types representative of all three germ layers, can form embryoid bodies, and localize in the ICM when injected into blastocysts [10, 11]. Importantly, we have produced chimeric pigs using blastocyst injection at relatively high efficiencies [10], and are currently analyzing progeny from these to determine germline transmission. This is only the second time that chimeric pigs have been reported in the peer review literature [12]. Furthermore, we have demonstrated that this method can be used to isolate homogeneous outgrowths in mice as well as cattle (unpublished results), suggesting that our method may facilitate isolation of ESCs from a range of species including those for which ESCs are yet to be isolated. The isolation of porcine ESCs will allow a range of biomedical as well as agricultural applications to be examined. In particular it will facilitate the production of multitransgenic pigs for xenotransplantation research, as well as provide a large animal model for human stem cell research [13].

2 Materials

2.1 Mice for Mouse Embryonic Fibroblast Feeder Layer

Pregnant 129X1/Sv JArc mice were used for the production of mouse embryonic fibroblast (MEF) feeder layers and were obtained from Animal Resource Centre (Canning Vale, Western Australia).

2.2 Pig Blastocyst Stage Embryos

In vitro and in vivo large white×landrace pig embryos were obtained as described [8, 10] with minor modifications. Namely, M199 oocyte maturation medium was supplemented with 5 μg/mL insulin, 7.5 μg/mL penicillin-G, and 50 μg/mL streptomycin sulfate.

2.3 Tissue Culture and Glassware

1. Center-well organ culture dishes.
2. 25 cm^2 Tissue culture flask.
3. 75 cm^2 Tissue culture flask.
4. 175 cm^2 Tissue culture flask.
5. 50-mL Syringe.
6. Ultrafine insulin syringe.
7. Glass Pasteur pipettes.
8. Glass capillaries (Clark Electromedical Instruments).
9. Hemocytometer.

Table 1
Composition of pESC medium and MEF medium

	pESC medium (50 mL)		MEF (500 mL)	
Reagent	**Final concentration**	**Amount**	**Final concentration**	**Amount**
DMEM	–	–	–	440 mL
FBS	–	–	10 %	50 mL
NEAA	1×	0.5 mL	1×	5 mL
PSF	1×	0.5 mL	1×	5 mL
α-MEM	–	44 mL	–	–
SR	10 %	5 mL	–	–
ITS	1×	0.5 mL	–	–
Sodium pyruvate	1×	0.5 mL	–	–
2-ME	1×	50 μL	–	–
bFGF	10 ng/mL	20 μL	–	–
hEGF	10 ng/mL	20 μL	–	–
Activin A	10 ng/mL	20 μL	–	–
hLIF	10 ng/mL	50 μL	–	–

10. Milex-GP, 0.22 μm filters.
11. 15-mL conical tube.
12. 50-mL conical tube.
13. Universal capillary holder (Eppendorf).
14. CVM Vitrification Kit (Cryologic, Australia) includes:

CVM Vitrification Block with Handle and Lid, CVM Fibreplugs and Sleeves (sterile), CVM Cryobath with Lid, CMV Cup for thawing.

2.4 Media and Solutions

2.4.1 Preparation of Growth Factors and Cytokine Stock Solutions

To prepare stock solutions of bFGF, hEGF, and activin A, dissolve 10 μg of each in 400 μL of α-MEM supplemented with 0.1 % of BSA and aliquot into 20 μL for storage at −20 °C. One aliquot of each is used for each 50 mL of pESC isolation and culture medium to obtain a final concentration of 10 ng/mL.

2.4.2 Media Used for the Isolation and Culture of Pig ESC (pESC Medium) and for Production of MEF Feeder Layer (MEF Medium)

The reagents used for the different culture media are listed in Table 1. 50 mL of pESC medium is prepared every 2 weeks or as required. 500 mL of MEF medium is prepared each month, or as required. After preparation, media are filtered through 0.22 μm filter and stored at 4 °C.

2.4.3 Vitrification and Warming Solution

The reagents used for preparation of vitrification and warming solution are listed below:

1. *α-MEM-HEPES medium.* 25 mM HEPES in α-MEM medium. Solution should be filtered, stored at 4 °C, and discarded after 1 week.
2. *ES-HEPES medium.* 20 % FCS in α-MEM-HEPES medium should be filtered, stored at 4 °C, and discarded after 1 week.
3. *1 M Sucrose solution.* 1 M sucrose in ES-HEPES medium. Solution should be filtered, stored at 4 °C and discard after 1 week.
4. *10% Vitrification solution.* 10 % Ethylene glycol and 10 % DMSO in ES-HEPES medium. Solution should be filtered, stored at 4 °C, and used on the day of preparation.
5. *20% Vitrification solution.* 20 % Ethylene glycol, 20 % DMSO, and 333 mM sucrose in ES-HEPES medium. Solution should be filtered, stored at 4 °C, and used on the day of preparation.
6. *Warming solution I.* 0.2 M sucrose solution in ES-HEPES medium.
7. *Warming solution II.* 0.1 M sucrose solution in ES-HEPES medium.

2.5 Preparation of Gelatinized Dishes

1. Pour 400 μL of 0.2 % gelatine into center-well organ culture dishes to cover the bottom. Keep dishes in laminar flow cabinet for 30–45 min.
2. Aspirate the excess of gelatine solution and transfer required amount of mitotically inactivated MEF (*see* **Note 4**).

3 Methods

3.1 Preparation of MEF Feeder Layer

1. MEF are isolated from 13.5 dpc 129X1/JArc fetuses as described [14] and trypsinized upon reaching confluence with 0.25 % trypsin–EDTA solution. After inactivation of trypsin with MEF medium, cells are centrifuged, the supernatant removed, the pellet resuspended in freezing medium (92 % FBS and 8 % DMSO), and 1 mL aliquots placed into cryovials and stored in liquid nitrogen.
2. Thaw the frozen vial in a 37 °C warm water bath and subsequently sterilize with 70 % ethanol.
3. Transfer the cell suspension in 15-mL conical tube, and slowly add 10 mL of MEF medium while shaking the tube. Centrifuge the cells at 300 × *g* for 5 min. Remove supernatant and resuspend cells in MEF medium.
4. Examine the amount of viable cells with trypan blue stain and transfer viable cells to 25 cm^2 flask (2.5×10^5 cells), 75 cm^2 flask (1×10^6 cells), or 175 cm^2 flask (2×10^6 cells).

5. Change medium the following day and culture cells until they reach confluence.
6. Harvest cells with trypsin–EDTA, neutralize trypsin with MEF medium, and centrifuge the cells at 300 × *g* for 5 min. Remove supernatant and resuspend cells in MEF medium.
7. Mitotically inactivate MEF cells with gamma-irradiation at a total absorbed radiation of 30 Gy. Keep irradiated cells on ice before transfer to centre-well organ culture dishes.
8. Count cell amount per mL and transfer them into gelatinized centre-well organ culture dishes at cell concentration of $4.9–5 \times 10^4$ cells per cm^2. Leave dishes in laminar flow cabinet for 1 h for even distribution of cells over the dishes and then place them in CO_2 incubator at 37 °C.
9. MEF feeder layer could be used after 2 days of incubation (*see* **Notes 1–3**).

3.2 Isolation of Pig Pluripotent Primary Embryonal Outgrowth from In Vitro-Produced or In Vivo-Derived Blastocysts

1. Aspirate MEF medium from desired quantity of center-well organ culture dishes and pour 600–800 μL of pESC medium. Place them for 1 h in CO_2 incubator at 39 °C in an atmosphere of 90 % N_2, 5 % CO_2, and 5 % O_2 to equilibrate.
2. Use day-7 IVP blastocysts or day-5.5 in vivo-derived blastocysts. Blastocysts should be expanded, hatched, or hatching. Embryos should have prominent ICM, large blastocoele cavity, and large number of trophoblast cells (*see* **Note 5**).
3. Transfer blastocysts (Fig. 1a) to Tyrode's solution for 2–3 min to remove zona pellucidae. Using a stereomicroscope watch when zona pellucida starts thinning. Transfer blastocysts to ESC medium, and gently pipette these using a Pasteur pipette to remove the zonae.
4. Gently squash the zona pellucida-free blastocysts by repeated pipetting using a Pasteur pipette whose tip diameter is about half that of the blastocysts.
5. Transfer squashed blastocysts into dishes with equilibrated pESC medium (up to ten per dish).
6. Using ultrafine insulin syringe attached to a 29-Gauge needle, gently press and flatten squashed blastocysts onto feeder layer (Fig. 1b). Place the dishes with plated blastocysts back in the CO_2 incubator, and culture them at 39 °C in humidified atmosphere of 90 % N_2, 5 % CO_2, and 5 % O_2.
7. Observe the plated embryos after 2 days. The majority of trophoblast cells will start to detach from feeder layer and degenerate, leaving clusters of few undifferentiated ICM cells attached to feeder layer (Fig. 2a). Over the next 3–5 days trophoblast cells degenerate and are eliminated from the culture. Small primary colonies composed of cells derived from undifferentiated ICM begin to form around the same time.

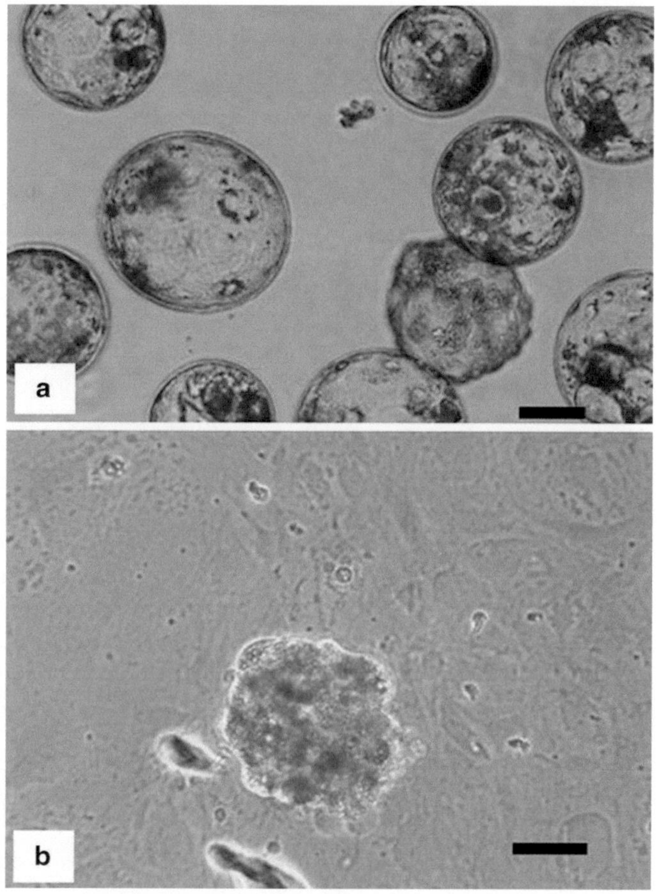

Fig. 1 In vitro-produced day-7 pig embryos (**a**) or in vivo-derived day-5.5 pig embryos after removal of zonae pellucidae were gently pressed onto mitotically inactivated MEF (**b**), to establish homogeneous primary embryonal outgrowths. Scale bars: **a**—200 μm; **b**—75 μm

8. Identify colonies with morphology of cells characteristic of ESC: Cells should have polygonal shape, relatively small (12–15 μm) diameter, a high nuclear/cytoplasmic ratio, and multiple lipid inclusion. Colonies should grow as monolayers without the presence of any extraembryonic endoderm cells.
9. Let the colonies reach 2–5 mm in diameter (Fig. 2b) for another 4–8 days changing medium every second day.

3.3 Passaging of Primary ESC Colonies

1. One day before passaging change the pESC medium to the dishes.
2. Pull a glass capillary using gas burner and use the fine polishing stone to cut the pulled capillary to create a sharp blunt tip. Insert capillary into universal capillary holder and use it to cut primary pESC colonies for mechanical passaging.

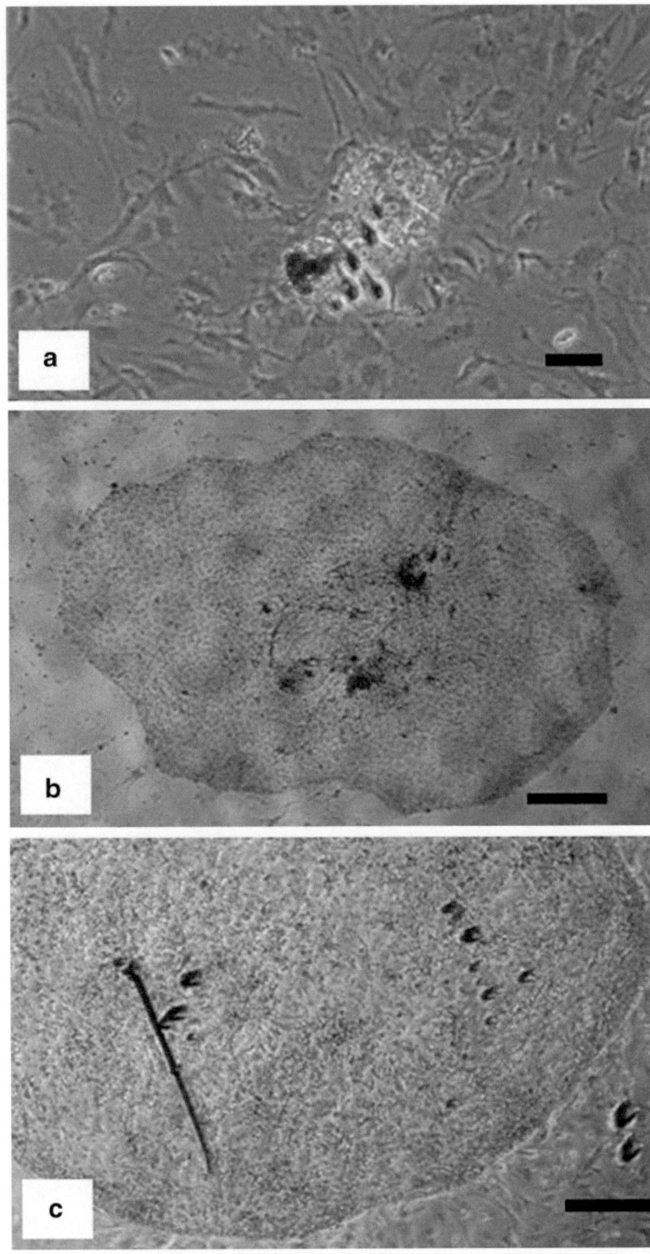

Fig. 2 1–2 days after plating the majority of trophoblast cells start to detach from MEF whereas undifferentiated ICM cells remain attached to feeder layer (**a**). These cells continue to grow for another 8–12 days forming homogeneous primary embryonal outgrowths (**b**) whose cells display a morphology characteristic of ESCs. After mechanical passaging these homogeneous primary outgrowths give rise to homogeneous pig ESC lines (**c**). Scale bars: **a**—150 μm; **b**—400 μm; **c**—500 μm

3. Cut primary colonies of pESC onto small pieces slightly touching the colonies with the sharp tip of capillary. Cutting needs to be done very gently to avoid dislodging colonies from feeder layer.

4. Transfer colony pieces (up to ten) onto dishes with fresh feeder layer.
5. Follow the same procedure as in Subheading 3.2, **step 6** above, to plate colony pieces onto feeder layer.
6. After 5–7 days of culture, depending on the size of cut pieces, homogeneous colonies of pluripotent cells (Fig. 2c) will appear.
7. Repeat the same procedure as in **steps 1–6**.
8. Vitrify colonies as described in Subheading 3.4 every two to three passages to avoid loss of ESC lines.

3.4 Cryologic Vitrification of pESC

1. Pour the following solutions into the four center-well vitrification dishes: dish 1–0.7 mL of ES-HEPES medium; dish 2–0.7 mL of ES-HEPES medium; dish 3–0.7 mL of 10 % vitrification solution; dish 4–0.7 mL of 10 % vitrification solution. Place dishes in CO_2 incubator for equilibration for 45–60 min.
2. Cool down vitrification block in cryobath with liquid nitrogen.
3. Label an appropriate number of fingerplugs and prepare and label cane.
4. Top up the cryobath with liquid nitrogen after vitrification block has cooled.
5. Cut colonies as described in Subheading 3.3.
6. Put all pieces of the colonies to be cryopreserved into dish 1.
7. Move only the number of pieces to be cryopreserved on one cane into dish 2.
8. Move the pieces to be placed onto one fiberplug (up to ten) into dish 3 (10 % vitrification solution), wash briefly, and leave for a total of 60 s.
9. Transfer pieces into dish 4 (20 % vitrification solution), wash quickly but thoroughly, and then suck up into pipette in the smallest possible volume.
10. Load the pieces onto the hook of the fiberplug in as small a droplet as possible.
11. Carefully touch the droplet onto the precooled vitrification block surface, and cover it with precooled sleeve. Hold the join between them briefly to slightly warm it before pushing them tightly together. Keep sleeve-covered fiberplug in precooled holding hole.
12. Repeat **steps 8–11** until all pieces are vitrified.
13. After all the pieces have been vitrified, cool down the cane, load the fiberplugs, place the cap over the fiberplugs, and then place cane into Dewar with liquid nitrogen (*see* **Notes 6** and **7**).

3.5 Warming pESC

1. Pour the following solutions into four center-well dishes: dish 1–0.7 mL warming solution I; dish 2–0.7 mL warming solution II; dish 3–0.7 mL ES-HEPES medium; dish 4–0.7 mL ES-HEPES medium.
2. Recover the fiberplugs to be thawed from the Dewar and keep them in a cup for thawing filled with liquid nitrogen.
3. When ready to thaw, grasp the fiberplug from the liquid nitrogen with forceps, hold the handle by the top, gently warm the fiberplug/sleeve join, and then loosen the join with a gentle twisting action while keeping the hook emerged into liquid nitrogen.
4. Quickly move the fiberplug from the liquid nitrogen and stir the vitrified droplet in dish 1 until it has dissolved.
5. Wash the pieces of colonies in dish 1 for about 1 min.
6. Transfer into dish 2 and leave for 5 min.
7. Transfer into dish 3 and leave for 5 min.
8. Transfer into dish 4 and leave for 5 min.
9. Transfer the pieces of colonies (up to ten) to dishes with fresh feeder layer.
10. Follow the same procedure as in Subheading 3.2, **step 6** above, to plate the pieces of colonies onto feeder layer.

4 Notes

1. Remember to fill up the humidification chamber around center well with 2–2.5 mL of sterilized culture-grade water to prevent drying out the center well.
2. The amount of inactivated MEF cells plated on cm^2 of center well dish is critical. If not enough MEF cells were plated, the colonies of pESC will start differentiate. If cell density were high, the MEF cells will be pushed out of feeder layer and also lug away the growing colonies.
3. Do not use feeder layers older than 2 weeks.
4. During the dish gelatinization do not allow the gelatine to dry and plate inactivated MEF immediately after aspiration of gelatine.
5. The stage of development is critical for isolation of homogeneous pluripotent primary embryonal outgrowths. The blastocysts should be at a stage when ICM is yet to differentiate into epiblast and hypoblast. The optimal stage for pig in vivo-derived embryos is days 5.0–5.5.
6. Use at least three pipettes, one for transfer from dish 1 to dish 2, from dish 2 to dish 3, and from dish 3 to dish 4 and onto the fiberplug.

7. Place an appropriate number of fiberplug sleeves on the holding holes on the vitrification block before starting vitrification. Make sure that sleeves have cooled sufficiently before covering the fiberplugs.

References

1. Tachibana M, Sparman M, Ramsey K et al (2012) Generation of chimeric rhesus monkeys. Cell 148:285–295
2. Hanna JH, Saha R, Jaenisch R (2010) Pluripotency and cellular reprogramming: facts, hypothesis, unresolved issues. Cell 143:508–524
3. Evans MJ, Kaufman MH (1981) Establishment in culture of pluripotential cells from mouse embryos. Nature 292:154–156
4. Martin GR (1981) Isolation of pluripotent cell line from early mouse embryos cultured in medium conditioned by teratocarcinoma stem cells. Proc Natl Acad Sci USA 78:7634–7638
5. Keefer CL, Pant D, Blomberg L et al (2007) Challenges and prospects for the establishment of embryonic stem cell lines of domesticated ungulates. Anim Reprod Sci 98:147–168
6. Nowak-Imialek M, Kues W, Carnwath JW et al (2011) Pluripotent stem cells and reprogrammed cells in farm animals. Microsc Microanal,doi:10.1017/s1431927611000080
7. Vasil'eva S, Vasil'ev I (1995) The effect of the developmental stage and of the cultivation conditions for cow and mouse embryos on the isolation of ES-like cells. Ontogenez 26:206–212
8. Vassiliev I, Vassilieva S, Beebe LFS et al (2010) Development of culture conditions for the isolation of pluripotent porcine embryonal outgrowths from in vitro produced and in vivo derived embryos. J Reprod Dev 56:546–551
9. Talbot NC, Garrett WV (2001) Ultrastructure of the embryonic stem cells of the 8-day pig blastocyst before and after in vitro manipulation: development of junctional apparatus and the lethal effects of PBS mediated cell-cell dissociation. Anat Rec 264:101–113
10. Vassiliev I, Vassilieva S, Beebe LFS et al (2010) In vitro and in vivo characterization of putative porcine embryonic stem cells. Cell Reprogram 12:223–230
11. Vassiliev I, Vassilieva S, Truong KP et al (2011) Isolation and in vitro characterization of putative porcine embryonic stem cells from cloned embryos treated with trichostatin A. Cell Reprogram 13:205–213
12. Chen LR, Shiue YL, Bertolini L et al (1999) Establishment of pluripotent cell lines from porcine preimplantation embryos. Theriogenology 52:195–211
13. Nottle MB, Vassiliev I, O'Connel PJ et al (2010) On the need for porcine embryonic stem cells to produce Gal KO pigs expressing multiple transgenes to advance xenotransplantation research. Xenotransplantation 17:411–412
14. Robertson EJ (1987) Embryo derived stem cells. In: Robertson EJ et al (eds) Teratocarcinomas and embryonic stem cells: a practical approach. IRL, Oxford, pp 19–49

Chapter 8

Isolation and Culture of Pig Epiblast Stem Cells

Aida Rodriguez, David A. Contreras, and Ramiro Alberio

Abstract

Regulation of early development can be directly interrogated in the embryo or can be studied in cultured cells isolated in the laboratory. New understanding of the developmental stages and the signalling requirements of the pig embryo have enabled the development of improved protocols for the derivation of pluripotent cells from early epiblasts. Here, we provide a detailed step-by-step description of the critical parameters required for isolation and establishment of these cells and how they can be used to study their developmental properties.

Key words Pig epiblast, Pluripotency, Epiblast stem cell, Activin/nodal signalling

1 Introduction

Early mammalian development ensues through hierarchical specification of cells. Two successive differentiation events result in the segregation of three cell lineages that together form the blastocyst [1]. The first segregation occurs at the compact morula stage with the outer layer of cells forming the epithelial trophectoderm (TE), which develops into the embryonic part of the placenta, and the inner layer of cells which form the inner cell mass (ICM). The second segregation divides the ICM into the epiblast and hypoblast, which give rise to the embryo proper and to the yolk sac, respectively. These three cell populations have very distinctive transcriptional programs that define their developmental capacity [1].

Embryonic stem cells (ESC) are derived from the early epiblast of mouse blastocysts [2]. These cells have the ability to differentiate into all embryonic tissues, grow indefinitely in culture, and can efficiently contribute to germline chimeras. The signalling pathways involved in ESC pluripotency have been extensively studied in the last decades establishing that leukemia inhibitory factor (LIF) and BMP4 are essential factors for maintaining their undifferentiated state [3, 4]. Establishment of ESC from other mammalian species using the conditions developed in mice has proven more

Ramiro Alberio (ed.), *Epiblast Stem Cells: Methods and Protocols*, Methods in Molecular Biology, vol. 1074,
DOI 10.1007/978-1-62703-628-3_8, © Springer Science+Business Media New York 2013

difficult, and this has been attributed, in part, to the different requirements of pluripotent cells in different species. Indeed, stem cells derived from postimplantation (5.5 days) mouse epiblast are dependent on FGF2 and activin/nodal signalling [5, 6] for their undifferentiated proliferation. Similar signalling requirements have been determined for human [7], rabbit [8], and pig [9] pluripotent cells derived from early embryos, indicating that pluripotency can be captured under different conditions depending on the species. Although mouse ESC and EpiSC share many stem cell features, including differentiation capacity into multiple somatic lineages, only ESC are capable to contribute to germline chimeras efficiently. There are no reports on germline chimerism in farm animal species after injection of stem cells into blastocysts, which raises the question of whether FGF/Activin A-dependent pluripotent cells have equivalent in vivo developmental capacity to LIF-dependent stem cells. Attempts to generate LIF-dependent pig pluripotent cells (or naïve cells) from various sources (blastocysts and induced pluripotent cells) using genetic manipulation have been successful [10, 11]; however, a report investigating the in vivo differentiation capacity of LIF-dependent pig iPS cells shows that these cells cannot colonize the germline [12].

This new knowledge on the differences in the regulation of pluripotency in mammals highlights the importance of conducting comparative studies to increase our basic understanding of embryo development. We hope that the protocols presented in this chapter are useful to scientists using the pig as a research animal model and enable the development of genetic manipulation technologies using stem cells.

2 Materials

2.1 Cell Culture Media and Buffers

2.1.1 Mouse Fibroblast Isolation and Preparation of Feeder Cells

1. Phosphate buffered saline (PBS) Ca^{2+}/Mg^{2+} free, adjusted to pH 7.4 and sterilized by autoclaving. Store at 4 °C.
2. PBS + Pen/Strep: PBS Ca^{2+}/Mg^{2+} free, penicillin 100 units/mL, and streptomycin 100 μg/mL. Store at 4 °C for up to 2 weeks.
3. 0.25 % Trypsin–EDTA.
4. MEF medium: Dulbecco's Modified Eagle's Medium (DMEM) 4,500 mg/L glucose, 10 % fetal calf serum (FCS), 2 mM L-glutamine, penicillin 100 units/mL and streptomycin 100 μg/mL, 1 % minimal essential medium (MEM) nonessential amino acids, and 0.1 mM β-mercaptoethanol. Filter medium through 0.22 μm pore syringe filter. Store at 4 °C for up to 2 weeks.
5. Mitomycin C stock: dissolve mitomycin C to 100 mM in DMSO. Long-term storage at −20 °C.

6. Mitomycin C working solution: sterile DMEM 4500 mg/L glucose, 10 % FCS, 2 mM L-glutamine, penicillin 100 units/mL and streptomycin 100 μg/mL, 10 μg/mL Mitomycin C. To help Mitomycin C solubility, add pre-warmed medium.
7. Porcine skin gelatine solution: 0.1 % (w/v) gelatine in distilled H_2O. Dissolve gelatine in water solution at 50 °C by stirring. Sterilize by autoclaving. Store at 4 °C for up to 4 weeks.
8. Freezing solution: 20 % (v/v) dimethyl sulfoxide in filtered FCS. Store at 4 °C for a week (*see* **Note 1**).
9. 70 % Ethanol (*see* **Note 2**).

2.1.2 Pig Embryo Collection, Epiblast Isolation and Culture

1. Flushing buffer: sterile PBS Ca^{2+}/Mg^{2+} free, penicillin 100 units/mL and streptomycin 100 μg/mL, 1 % FCS. Store at 4 °C for up to 1 week.
2. Washing medium: DMEM/F12, 20 % FCS, penicillin 100 units/mL, and streptomycin 100 μg/mL. Filter medium through 0.22 μm pore syringe filter. Store at 4 °C for up to 2 weeks.
3. Holding Medium: DMEM/F12, 20 % FCS, 1 % MEM nonessential amino acids, 1 % glutamine, penicillin 100 units/mL and streptomycin 100 μg/mL, and 0.1 mM β-mercaptoethanol, 25 mM HEPES (*N*-2-Hydroxyethylpiperazine-*N*-2-Ethane Sulfonic Acid) buffer. Filter medium through 0.22 μm pore syringe filter. Store at 4 °C for up to 2 weeks.
4. Basic Fibroblast Growth Factor (bFGF) stock: reconstitute at 100 μg/mL in sterile PBS 0.1 % (w/v) Bovine Serum Albumin (BSA). Store at −20 °C up to 12 months.
5. Epiblast and EpiSC (epiblast stem cells) culture medium: DMEM/F12, 1 % MEM nonessential amino acids, 1 % glutamine, penicillin 100 units/mL and streptomycin 100 μg/mL, and 0.1 mM β-mercaptoethanol, either 20 % FCS or Knockout serum replacement (KSR) and bFGF 5 ng/mL.
6. Enzymatic solution: 1 mg/mL Collagenase IV and 1 mg/mL Dispase in DMEM/F12. Filter solution through 0.22 μm pore syringe filter and keep at −20 °C for long-term storage.
7. *Vitrification solutions*:
   (a) Holding Medium: (detailed above).
   (b) VS1: 10 % (v/v) dimethyl sulfoxide (DMSO) and 10 % (v/v) ethylene glycol (EG) in Holding Medium.
   (c) VS2: 20 % (v/v) DMSO, 20 % (v/v) EG and 17.1 % (w/v) sucrose in Holding Medium.
   (d) Prepare fresh solutions, filter through 0.22 μm pore syringe filter and storage at −4 °C. Use within 1–2 days.

8. *Thawing solutions*:
   (a) Holding Medium: (detailed above **step 7(a)**).
   (b) TS1: 6.85 % (w/v) sucrose in Holding Medium.
   (c) TS2: 50 % (v/v) TS1, 50 % (v/v) Holding Medium.

Prepare fresh solutions, filter through 0.22 μm pore syringe filter and storage at –4 °C. Use within 1–2 days.

9. *Solutions for pig EpiSC Immunocytochemistry (ICC)*:
   (a) Fixative: Paraformaldehyde (PFA) 4 %: Solution A: 2 g $NaH_2PO_4{\cdot}H_2O$ in 100 mL distilled water. Solution B: 2.5 g NaOH in 100 ml distilled water.
   (b) Mix 8.3 mL of solution A with 1.7 mL of solution B and add 400 mg of PFA. To dissolve the PFA, stir and heat the solution up to 60 °C for 2 h. Adjust to pH 7.4 (*see* **Note 3**).
   (c) Washing buffer: 1 % BSA (w/v) in PBS.
   (d) Permeabilization solution: 0.1 % (v/v) Triton X-100 in Washing buffer.
   (e) Blocking solution: 5 % BSA (w/v) in PBS.
   (f) DAPI: 1 μg/mL solution in Washing buffer.
   (g) Mounting medium: Aqueous mounting.
10. *Karyotyping solutions*:
   (a) Demecolcine: Demecolcine solution is stored at 4 °C and is used at a final concentration of 0.05 μg/mL.
   (b) Fixative: Acetic acid:methanol (1:3) solution. Keep at –20 °C.
   (c) KCl solution: 0.075 M (56 mg/10 mL) KCl in distilled water. Store at 4 °C and warm up to 37 °C before use.
   (d) Gurr Buffer. Dissolve Gurr tablet in distilled water. Stir until fully dissolved and keep at RT.
   (e) Giemsa staining solution: KaryoMAX Giemsa Stain Stock Solution in Gurr Buffer (5:50). Stir until fully dissolved and keep at RT.
11. *Alkaline phosphatase (AP) activity, colorimetric assay* (Sigma AP detection kit):
   (a) Fixative: 6.25 mL citrate solution (provided by the kit), 16.25 mL acetone, 2 mL formaldehyde 37 %. Mix by inversion and keep refrigerated (*see* **Note 4**).
   (b) AP assay mixture: mix by inversion 100 μL sodium nitrite solution (solution A) and 100 μL Fast Red Violet alkaline solution (solution B). Allow the mixture to stand for 2 min. Add 4.5 mL of deionized water, 100 μL Naphthol AS-BI alkaline solution and mix thoroughly. Use the mixture within 30 min.

2.2 Equipment

1. Portable incubator.
2. Stereomicroscope.
3. Emcon filter.
4. Scissors.
5. Bacteriological dishes (90 and 35 mm).
6. Sterile 20 mL syringes.
7. Surgical blades.
8. Cryogenic vials (2 mL).
9. Glass dishes 90 mm.
10. Needles 21 G (0.8 × 40 mm) and needles 18 G.
11. Fine stainless steel forceps (Biology Nr. 4).
12. Micropipettes.
13. Disposable sterile 2 mL plastic pipettes.
14. Disposable pipettes: 5, 10, and 25 mL.
15. Cell culture plates (4 well and 6 well).
16. Glass coverslips (round, 11 mm diameter).
17. Glass slides.
18. Polypropylene tubes (50, 15 mL).
19. Open Pulled Straws (OPS).

3 Methods

3.1 Mouse Feeder Cells Preparation

3.1.1 Mouse Fetal Fibroblast Isolation and Culture

1. Sacrifice E13.5 pregnant mice by cervical dislocation and transfer them to a laminar flow hood.
2. Clean the abdomen with 70 % ethanol, make an initial incision in the skin and tear open.
3. Cut open peritoneum with scissors to expose uterine horns. Dissect out the uterine horns and immerse briefly in 70 % ethanol.
4. Place uterine horns in a petri dish with sterile PBS.
5. Separate the fetuses from the decidua and transfer to a new dish with PBS (*see* **Note 5**).
6. Remove the fetuses from the amnion, decapitate each one as quickly as possible, and placed them in a fresh petri dish containing PBS.
7. Remove "red" organs (liver and heart), and as much blood as possible.
8. Wash the fetuses twice in PBS + Pen/Strep.

9. Place pooled fetuses in 1 mL 0.25 % Trypsin–EDTA in a small dish and mince the tissues with surgical blades. When the mixture is free of large pieces of tissue, transfer to 50 mL centrifuge tube and add trypsin (up to 2 mL per fetus). Use syringe and 21 G needle to completely triturate the fetuses.
10. Vortex 20 s and incubate at 37 °C for 5 min. Repeat this step until the number of large clumps is reduced.
11. To inactivate the trypsin, add 3 mL MEF medium per fetus and shake vigorously. Let clumps settle down (*see* **Note 6**).
12. Place the fibroblast into T175 flasks at a density of two fetuses per flask and top up with 30 mL MEF medium (*see* **Note 7**). This is the passage 0 (P0). Incubate at 37 °C and 5 % CO_2.
13. Change to fresh medium on the following day.
14. Once the flasks become confluent, remove the MEF medium and wash twice with PBS.
15. Add pre-warmed 0.25 % Trypsin–EDTA (enough volume to cover the cells) and disperse by gently tilting the flask. Incubate at 37 °C for ~2–3 min.
16. When the cells detach from the flask, inactivate the trypsin by adding MEF medium (about 10 mL of MEF medium per 1 mL of trypsin), and pipette up and down a few times to dissociate MEFs into single cells.
17. Pellet the cells 300 × *g* for 5 min and remove supernatant.
18. Count the cells and freeze (*see* Subheading 3.1.3 Feeders cryopreservation).

3.1.2 Fibroblast Inactivation and Preparation of Feeder Cells

1. When MEFs reach ~90 % confluence, split them to passage 3 (P3) at a split ratio of 1:3 (*see* **Note 8**).
2. Mitomycin-inactivate MEFs when they reach 90 % confluence at P3.
3. Remove MEF medium and add Mitomycin C working solution (10 mL to a 75 cm^2 flask).
4. Incubate MEF with Mitomycin 3 h at 37 °C and 5%CO_2 (*see* **Note 9**).
5. Remove the Mitomycin C and wash twice with PBS.
6. Trypsin the inactivated MEF following step previously described (Subheading 3.1.1, **steps 15–17**).
7. Freeze the inactivated MEF (*see* Subheading 3.1.3) or plate them as a feeder cells.
8. To use the inactivated MEFs as a feeder layer, plate them onto gelatine-coated wells (*see* **Note 10**) at a density of 30,000–40,000 cells/cm^2 in MEF medium.
9. Incubate at 37 °C and 5 % CO_2 overnight before using (*see* **Note 11**).

3.1.3 Fibroblast Cryopreservation and Thawing

Cryopreservation

1. Pellet the MEFs at 300 × *g* for 5 min and count the cell number.
2. Dilute the MEFs pellet at a desired concentration with MEF medium and gently resuspend the pellet. Aliquot 0.5 mL of MEFs suspension in sterile cryovials labelled with the passage number, date, and concentration.
3. Top up the MEF suspension with equal volume of freezing solution (0.5 mL), add the freezing solution drop-wise.
4. Gently shake the cryovial and quickly transfer into a cryo-container and store more than 15 h at −80 °C.
5. The following day transfer the cryovials into a liquid nitrogen tank.

Thawing

1. Take out the vials from the liquid nitrogen and thaw quickly in a water bath pre-warmed at 35–37 °C.
2. Transfer the cryovial content to a tube with 10 mL of pre-warmed MEF medium. Pellet by centrifugation at 300 × *g* for 10 min (*see* **Note 12**).
3. Remove the supernatant and resuspend the pellet in MEF medium.

3.2 Pig Embryo Collection and Epiblast Isolation and Culture

3.2.1 Artificial Insemination and Embryo Collection

1. Sows are artificially inseminated twice (12 h apart) at the beginning of estrus. Typically, 72 h after weaning sows show signs of estrus and are ready for insemination. Ten to eleven days after artificial insemination the sows are slaughtered and the uterine tracts are retrieved within 30 min.
2. Uterine horns are freed from the mesometrium with scissors (Fig. 1a) and a clamp is placed near the cervix before cutting this end.
3. A 20 mL syringe loaded with warm flushing buffer is emptied in the oviducts to flush the uterine content by applying a gentle massage.
4. The flush is collected into an EmCon embryo filter (*see* **Note 13**).
5. Embryos are macroscopically (Fig. 1b) visible and can be transferred to a dish for washing, before transfer to a holding medium to be transported to the laboratory at 38 °C.

3.2.2 Epiblast Isolation and Initial Culture

1. Transfer embryos to a dish containing fresh holding medium for dissection.
2. Using fine biology forceps (Nbr. 4) and needles (18 G) the epiblasts are gently separated from remnants of the Rauber's layer, the surrounding trophoblast, and the hypoblast (Fig. 1c–f).
3. Dissected epiblasts (Fig. 2a) are washed twice in washing medium.

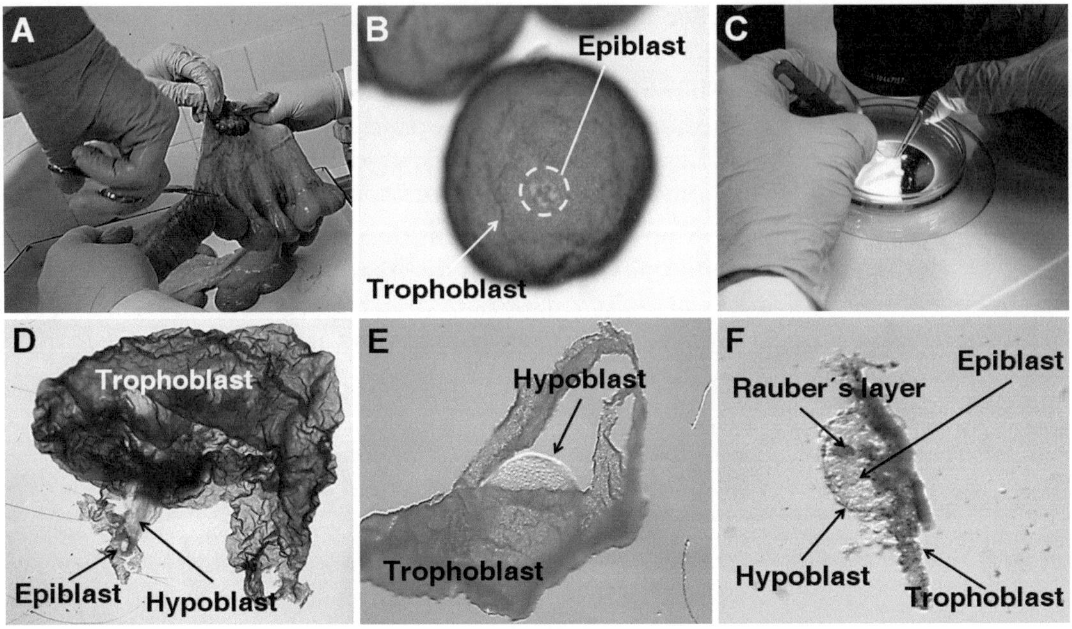

Fig. 1 (**a**) Genital tract is transported to the laboratory and the uterine horns are released cutting out the mesometrium to facilitate embryo flushing. (**b**) Pre-streak stage I embryo. Exposed epiblast surrounded by trophoblast. Note Rauber's layer covering the epiblast. (**c**) Epiblast isolation procedure under stereomicroscope, helped with tweezers. (**d**) Once the Epiblast has been localized, tear open the trophoblast and the hypoblast will be exposed together with the Epiblast. (**e**) Intact hypoblast at pre-streak stage I shows a spherical shape. (**f**) Epiblast after isolation with some cells from different tissue of origin attached: hypoblast, Rauber's layer, and trophoblast

4. Transfer the epiblasts into single wells of a 4-well plate containing Epiblast stem cell medium and inactivated MEFs (*see* **Note 14**).
5. Epiblasts are cultured for 5–10 days until they reached 5–7 times their original diameter (*see* **Note 15**) (Fig. 2e).

3.2.3 Epiblast Stem Cells Culture and Colony Passages

1. When the colonies reach the optimal size, groups of cells with high nuclear/cytoplasmic ratio can be identified. These should be cut out using a stem cell knife or a fine needle (*see* **Note 16**).
2. The small clusters are transferred to 3.5 cm dishes containing inactivated MEFs.
3. After overnight incubation, good quality cultures will display attached colonies and after 24 h they will show signs of cell proliferation. At this stage and for the next two passages there will still be some clumps differentiating to extraembryonic cell types and should be removed from the cultures.

3.2.4 EpiSC Cryopreservation: Vitrification and Thawing

Prepare the material and pre-warm the solutions that are required for this procedure before you take out the colonies to be vitrified. Preparation of OPS is described in **Note 17**.

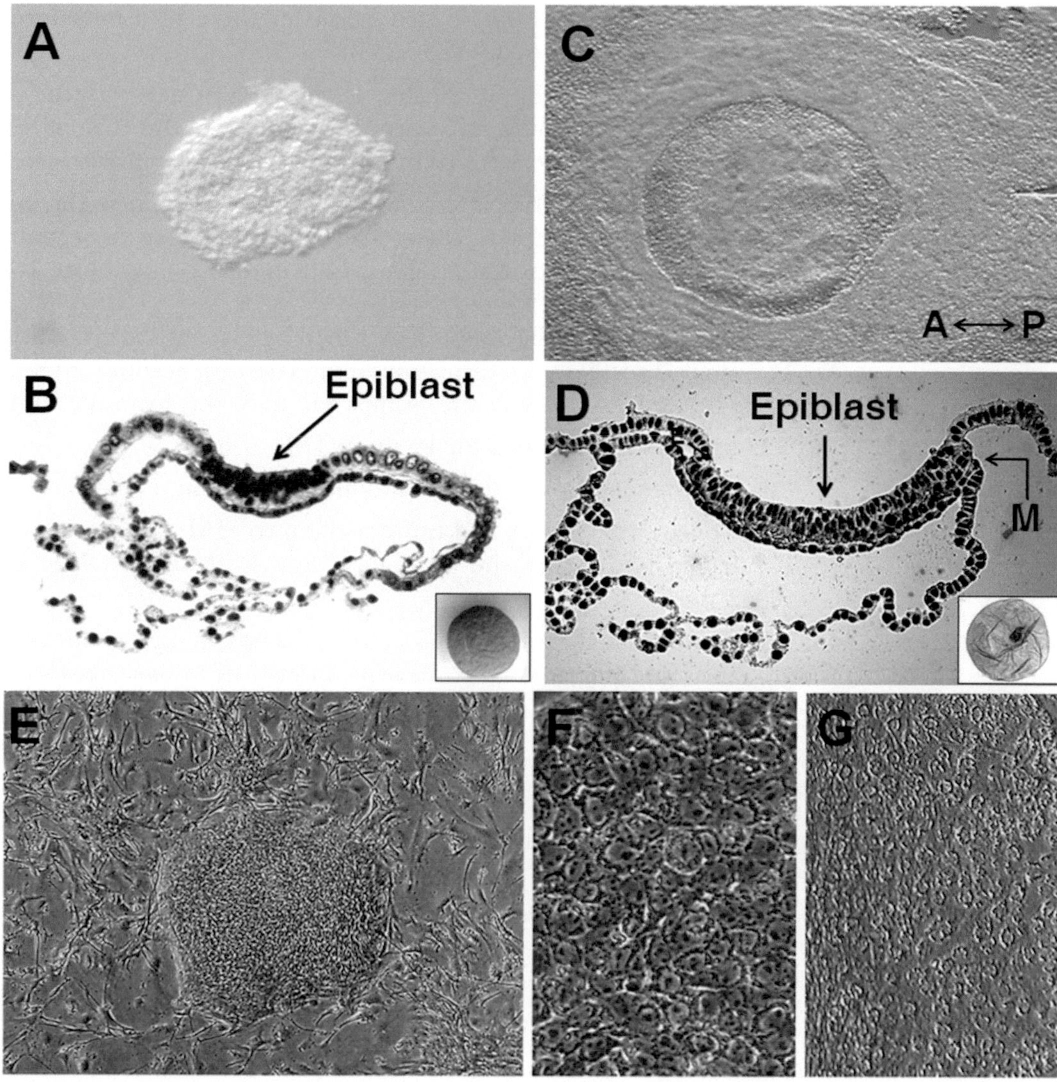

Fig. 2 (**a**) Top view of dissected epiblast. (**b**) Section of pre-streak stage I embryo. (**c**) Top view of pre-streak II epiblasts. (**d**) Section of pre-streak stage II embryo. (**e**) EpiSC colony, 5–6 days after plating. (**f**) EpiSC colony core ((**e**) at higher magnification). EpiSC show prominent nucleoli and high nuclear/cytoplasmic ratio. (**g**) Trophoblast-like differentiated cells

Vitrification

1. When the colonies reach the appropriate size, cut out into pieces following the steps mentioned above (Subheading 3.2.3, **step 1** and *see* **Note 16**).
2. Put about 4–6 clusters in a dish with pre-warmed Holding Medium (*see* **Note 18**).
3. Take about 50 μL of VS1 in a tip, collect the EpiSC clusters, transfer them to a well with VS1, and incubate 1 min.
4. Take about 50 μL of VS2 in a tip, collect the clusters from VS1, and transfer them to a dish/well with VS2. Incubate for 20 s.

Take the clumps on VS2 and transfer them to a microdrop (*see* **Note 19**).

5. To load the pig EpiSC clusters in the OPS, touch with the tip of the OPS the VS2 microdrop, the drop and the clusters will be loaded (about 4–5 μL) inside the straw by capillarity effect.
6. Submerge the OPS into liquid nitrogen (LN_2), first the end containing the EpiSC clusters and after a couple of seconds, when the VS2 is frozen, drop the OPS in the LN_2 (*see* **Note 20**).

Thawing

1. The pig EpiSC thawing is performed by placing the end of the straw directly into a dish containing Thawing Solution (TS) 1. Within a few seconds the clumps will thaw out from the straw.
2. Incubate the clusters for 1 min in TS1.
3. Transfer the EpiSC to TS2 and incubate 5 min.
4. Collect the clusters and transfer them to HM. Incubate them 5 min and repeat the incubation in a fresh dish of HM for 5 min.
5. Wash once the pig EpiSC colony clumps in EpiSC medium and plate them.

3.2.5 EpiSC Characterization: ICC, Karyotyping, and AP Assay

Immunocytochemistry

1. To perform the Immunocytochemistry (ICC), pig EpiSC are plated onto glass coverslips. When the colonies reach a good size (Fig. 2e) take the coverslip out of the well and wash once on ICC washing buffer, and fix with 4 % PFA for 20 min.
2. Wash three times in Washing buffer (5 min each). All the wash steps are performed on the rocker.
3. Incubate the cells in Permeabilization solution for 20 min (do not incubate in this solution the samples that are going to be stained for surface markers).
4. Wash twice in Washing buffer.
5. Block the samples for 1 h at RT in Blocking solution.
6. Dilute the Primary Antibodies at the desired concentration in Blocking solution. Incubate overnight at 4–8 °C.
7. Wash three times in Washing buffer (10 min each).
8. Incubate in the respective Secondary Antibody (diluted in Blocking solution) for 1 h at RT in the dark.
9. Wash more than three times in washing buffer (10 min each) in the dark.
10. Counter staining. Incubate the coverslip in DAPI solution (10 μg/mL) for 5 min in the dark.
11. Mount coverslip with a drop of mounting medium and check under microscope.
12. Seal the coverslip with nail polish and keep at −20 °C in dark.

Karyotyping

1. Remove the EpiSC medium and incubate the cells with Demecolcine (diluted in EpiSC medium) for about 40 min. For karyotyping, the cells need to be actively growing to increase the number of metaphases per slide.
2. Trypsinize cells for 1 min with 0.25 % Trypsin–EDTA and inactivate with 10 % FCS medium.
3. Pellet the cells by centrifugation at 300 × *g* for 5 min.
4. Remove the supernatant and resuspend the pellet in 400 μL KCl solution and mix by pipetting.
5. Incubate for 30 min at 37 °C.
6. Centrifuge 5 min at 300 × *g* and discard the supernatant.
7. Fix the pelleted cells with cold (−20 °C) fixative for 60 to 90 min at 4 °C (*see* **Note 21**).
8. Resuspend the pellet in 1 mL of cold fixative and add 9 mL more of fixative, mix the solution by inversion and keep it at RT for 5 min.
9. Centrifuge for 5 min at 300 × *g* and repeat **step 8** two more times.
10. After the last centrifugation, remove the supernatant and leave 1 mL to resuspend the cells (*see* **Note 22**).
11. Drop the cell suspension onto a clean humidified slide from about 30 cm of distance.
12. Dry the slides overnight at 37 °C.
13. Dehydrate the samples with EtOH. Incubate samples 2 min in 70 % EtOH, 2 min in 90 % and 2 min in 100 %.
14. Air-dry the slides.
15. Stain metaphase spreads with Giemsa staining solution for 20 min at RT.
16. Rinse the slides with water and allow them to dry at RT.
17. Mount the slides using mounting medium or 90 % glycerol.
18. Check under phase contrast using 100× objective. Count 20–30 metaphase spreads per cell line. Normal pig karyotype is $2N = 38$ chromosomes.

Alkaline Phosphatase Assay

1. Make up the Alkaline Phosphatase (AP) mixture fresh and warm up at RT. Keep the fixative at RT before use.
2. Remove the medium from the dish and wash once with PBS.
3. To fix the cells, add carefully the fixative and incubate for 1–2 min, remove the fixative and wash once with water.
4. Add the AP mixture and incubate for 20 min at RT in dark.
5. To stop the reaction, remove the AP mixture and wash twice with water. Add fresh water to the dish and check under microscope.

4 Notes

All the procedures involving animals have been approved by the School of Biosciences Ethics Review Committee (University of Nottingham, UK).

1. Although the freezing solution contains 20 % DMSO in FCS, the final concentration of DMSO in the cell suspension is 10 % DMSO.
2. To avoid contamination, clean the flow cabinet surface and the outside of the vials, pipettes, tip boxes with 70 % (v/v) ethanol.
3. Do not allow the PFA solution to boil, keep the mixture at <65 °C. Adjust to pH 7.4.
4. Warm up the fixative at RT prior to use.
5. Keep the fetuses from each female separate to make different pools of fetal fibroblasts.
6. The cell suspension should be free of any large pieces of tissue. Genomic DNA presence due to cell lysis increases the cell suspension viscosity, preventing the release of the single cells in the suspension. Addition of 100 Kunitz units DNAse I/mL reduces the viscosity.
7. Take 3–4 mL of cell suspension per fetus, avoiding the hydrophobic mass floating at the very top, and avoiding the tissue pieces at the bottom.
8. MEF provide a cellular matrix and secrete growth factors to support stem cells maintaining and growth. It is important to use MEFs at early passages to ensure good quality of feeders layers.
9. Mitomycin C reacts covalently with DNA forming crosslinks between the complementary strands of DNA preventing their separation and inhibiting DNA replication. Incubation periods longer than 3 h could decrease MEFs viability, and the quality of these cells as a feeder layers.
10. Coat the wells with porcine skin gelatine solution for 1 h at 37 °C. Before the inactivated MEFs are plated, remove the gelatine and let the well dry.
11. The feeder cells quality decreases with time. Do not use old feeder cells (plated earlier than 3 days before EpiSC are plated).
12. 10 min is the time needed by DMSO to exit the cells.
13. Most embryos are obtained in this first flush, but the procedure is repeated with another 20 mL to ensure that no embryos are lost.
14. A few hours prior to plating the epiblasts, remove the MEF medium from the plates, wash once with PBS, and add the EpiSC medium.

15. Because of the asynchrony of pig embryos in a single litter (Fig. 2a–d), epiblasts will differ significantly in size, and in the number of cells already committed to the hypoblast. This will result in significant differentiation of multiple parts of the plated epiblast from days 4 to 8 of culture. It is therefore critical to observe the cultures during this time and remove areas of overt differentiation. Typically these areas will form vesicles (mimicking the hypoblast) or will also spread as differentiated flat cells resembling trophoblast or extraembyonic mesoderm (Fig. 2g).
16. It is important not to make very small pieces, since this will affect the viability and promote the differentiation of these clumps.
17. OPS have been designed for rapid freezing (vitrification), to avoid ice crystal formation. To make the OPS, we use conventional 0.25 mL semen mini-straws (IMV, France). The mini-straws were held over a flame, until the straw become soft or slightly melted, and pulled manually up to twice its original length. Ideally, the inner diameter should decrease to half of the original diameter (from 1.7 to 0.8 mm), the wall thickness of the straw will decrease as well. After pulling, let the OPS cool down in air and cut with a sharp blade on a hard surface to obtain two OPS from one semen mini-straw with a diameter about 1–0.8 mm in the thinner tip.
18. It is important to keep the vitrification solutions (VS1 and VS2) warm, working in a heating stage set at 37–39 °C will help to avoid fluctuations in temperature.
19. The total incubation in VS2 should not exceed 25 s.
20. Before you drop the whole OPS into the LN_2, submerge the straw in the LN_2 and hold it until the air bubbles contained in the straws are released.
21. Fix the pellet until it becomes fully white.
22. The cells can be stored at –20 °C.

References

1. Arnold SJ, Robertson EJ (2009) Making a commitment: cell lineage allocation and axis patterning in the early mouse embryo. Nat Rev Mol Cell Biol 10(2):91–103, doi:nrm2618 [pii] 10.1038/nrm2618
2. Brook FA, Gardner RL (1997) The origin and efficient derivation of embryonic stem cells in the mouse. Proc Natl Acad Sci U S A 94(11): 5709–5712
3. Ying QL, Nichols J, Chambers I, Smith A (2003) BMP induction of Id proteins suppresses differentiation and sustains embryonic stem cell self-renewal in collaboration with STAT3. Cell 115(3):281–292, doi:S009286740300847X [pii]
4. Smith AG, Heath JK, Donaldson DD, Wong GG, Moreau J, Stahl M, Rogers D (1988) Inhibition of pluripotential embryonic stem cell differentiation by purified polypeptides. Nature 336(6200):688–690. doi:10.1038/336688a0
5. Brons IG, Smithers LE, Trotter MW, Rugg-Gunn P, Sun B, de Sousa C, Lopes SM, Howlett SK, Clarkson A, Ahrlund-Richter L,

Pedersen RA, Vallier L (2007) Derivation of pluripotent epiblast stem cells from mammalian embryos. Nature 448(7150):191–195, doi:nature05950 [pii] 10.1038/nature05950

6. Tesar PJ, Chenoweth JG, Brook FA, Davies TJ, Evans EP, Mack DL, Gardner RL, McKay RD (2007) New cell lines from mouse epiblast share defining features with human embryonic stem cells. Nature 448(7150):196–199, doi:nature05972 [pii] 10.1038/nature05972
7. Vallier L, Alexander M, Pedersen RA (2005) Activin/Nodal and FGF pathways cooperate to maintain pluripotency of human embryonic stem cells. J Cell Sci 118(Pt 19):4495–4509, doi:118/19/4495 [pii] 10.1242/jcs.02553
8. Honda A, Hirose M, Ogura A (2009) Basic FGF and Activin/Nodal but not LIF signaling sustain undifferentiated status of rabbit embryonic stem cells. Exp Cell Res 315(12):2033–2042, doi:S0014-4827(09)00049-4 [pii] 10.1016/j.yexcr.2009.01.024
9. Alberio R, Croxall N, Allegrucci C (2010) Pig epiblast stem cells depend on activin/nodal signaling for pluripotency and self-renewal. Stem Cells Dev 19(10):1627–1636. doi:10.1089/scd.2010.0012
10. Telugu BP, Ezashi T, Sinha S, Alexenko AP, Spate L, Prather RS, Roberts RM (2011) Leukemia inhibitory factor (LIF)-dependent, pluripotent stem cells established from inner cell mass of porcine embryos. J Biol Chem 286(33):28948–28953, doi:M111.229468 [pii] 10.1074/jbc.M111.229468
11. Rodríguez A, Allegrucci C, Alberio R (2012) Modulation of pluripotency in the porcine embryo and iPS cells. PLoS One 7(11), doi:10.1371/journal.pone.0049079
12. Fujishiro SH, Nakano K, Mizukami Y, Azami T, Arai Y, Matsunari H, Ishino R, Nishimura T, Watanabe M, Abe T, Furukawa Y, Umeyama K, Yamanaka S, Ema M, Nagashima H, Hanazono Y (2013) Generation of naive-like porcine-induced pluripotent stem cells capable of contributing to embryonic and fetal development. Stem Cells Dev 22(3):473–482. doi:10.1089/scd.2012.0173

Chapter 9

Isolation and Culture of Bovine Embryonic Stem Cells

Shanbo Cao, Fang Wang, and Lin Liu

Abstract

Isolation and culture of primary embryonic stem (ES) cell colonies are the first critical step towards establishment of stable ES cell lines. Here we introduce a novel method designated as "Separate and Seed" that contributes remarkably to efficient derivation of bovine primary ES cell colonies from blastocysts. The bovine ES cell colonies can self-renew to passage 10 with the growth factors bFGF and BIO. The bovine ES cells exhibit morphology typical of ES cells and express pluripotent molecular markers including Oct4, Nanog, SSEA1, SSEA4, and alkaline phosphatase (AP). These pluripotent markers may be used for the characterization of authentic bovine ES cell lines. Although continued efforts are required for improving long term culture of bovine ES cells, this novel "Separate and Seed" method plus the growth factors bFGF and BIO provides an initial effective step that may eventually lead to the derivation of authentic bovine ES cells.

Key words Bovine, Embryonic stem cells, Derivation, Separate and seed

1 Introduction

Embryonic stem (ES) cells are derived from preimplantation embryos. The first ES cell lines were isolated from mouse blastocysts in 1981 [1, 2], initiating a new era in the cell, developmental and reproductive biology, and regenerative medicine. ES cells exhibit two common characteristics: the ability to proliferate indefinitely while maintaining pluripotency in culture and the ability to differentiate into all derivatives of the three primary embryonic germ layers in vivo and in vitro [3–5]. In addition, rodent ES cells are able to contribute to germline transmissible chimeras [3].

Bovine ES cells are expected to have great implications in agricultural, pharmaceutical, and biomedical applications. Yet, it has been a tremendous challenge to establish bovine ES cell lines. There have been reports on derivation of bovine ES-like cells during the past two decades, but no "proven" stable truly pluripotent bovine ES cell lines are available [6–8]. These reported ES-like

Ramiro Alberio (ed.), *Epiblast Stem Cells: Methods and Protocols*, Methods in Molecular Biology, vol. 1074, DOI 10.1007/978-1-62703-628-3_9, © Springer Science+Business Media New York 2013

cells vary not only in morphology and specific ES molecular markers but also in chromosomal stability and pluripotency [9–19].

The pluripotency of ES cells is maintained by a combination of intrinsic and extrinsic factors that are either fundamental or species specific. Intrinsic factors involved in the regulation of pluripotency include Oct4, Sox2, and Nanog. Ectopic expression of Oct4 and Sox2, in combination with c-Myc and Klf-4 [20, 21] or with Nanog and Lin28 [22] can reprogram mouse and human somatic cells to pluripotent iPS cells. Then, similar factors have been used to establish reprogrammed iPS cells from several species such as monkey [23], rat [24], pig [25], goat [26], ovine [27], and bovine [28–30]. The reported bovine iPS are able to differentiate into derivatives of all the three basic germ layers in vivo and in vitro, and Nanog is reportedly a key factor for induction of pluripotency in bovine adult fibroblasts [30].

Extrinsic factors that maintain the pluripotency of ES cells include LIF/Stat3 [31–33] and BMP4/Smad [34, 35] pathways in mouse ES cells, and Activin/Nodal and bFGF signaling pathways in human ES cells [36, 37]. In addition, Wnt/β-catenin signaling pathway is one of the few common self-renewal pathways shared by both mouse and human ES cells [38, 39]. Activation of Wnt signaling pathway by 6-bromoindirubin-3′-oxime (BIO), a potent, reversible and ATP-competitive glycogen synthase kinase-3 (GSK-3) inhibitor, is shown to maintain self-renewal of both human and mouse ES cells that is possible through Wnt-induced expression of Nodal [40–42]. LIF and bFGF are the most commonly used factors for the maintenance of mouse and human ES cells, respectively. There are two LIF signaling pathways that are connected to the core circuitry via different transcription factors. Jak/Stat3 mainly activates Klf4 that preferentially triggers Sox2, and MAPK cascades regulate the activation of Tbx3 that predominantly stimulates Nanog [43]. bFGF signaling inhibits BMP4-induced differentiation of human ES cells by maintaining Nanog levels through the MEK–ERK pathway [44]. Comparatively, signaling required for maintenance of bovine ES cells remains elusive, and this might be one of the main reasons that no authentic bovine ES cell lines have been established thus far.

Isolation of primary ES cell colony is the first critical step in the establishment of stable ES cell lines, and the methods remain little changed during the past three decades. Most of mouse and human ES cell lines are generated directly from expanding blastocysts [45, 46], or ICM by immunosurgical treatment [47] or mechanical methods [48]. These approaches have been attempted in the derivation of bovine ES cells, but the efficiency and successful rates remain very low. Here, we introduce a novel high efficient method (up to 79.1 % successful rate), termed as "Separate and Seed," for the isolation of bovine primary ES colony [49]. The bovine ES cell colonies can self-renew to passage 10 with the growth factors bFGF and BIO. The ES cell colonies exhibit morphology typical of

ES cells and express specific pluripotent molecular markers such as Oct4, Nanog, AP, SSEA1, and SSEA4 that may be used for the indication of pluripotency and establishment of authentic bovine ES cell lines.

2 Materials

2.1 Feeder Cell Layers and Bovine Embryos

1. Feeder cell layers: mouse embryonic fibroblasts (MEFs) from 13.5-day-old B6D2F1 embryos.
2. Bovine embryos: in vivo produced or IVF-derived bovine mid-stage blastocysts.

2.2 Equipment and Tools

1. Vertical laminar flow bench.
2. Stereomicroscope.
3. Hot plate.
4. Inverted microscope.
5. Water-Jacketed CO_2 incubator.
6. Microcentrifuge.
7. Analytical balance.
8. 4-Well plates.
9. 35-mm Petri dishes.
10. 60-mm Petri dishes.
11. 100-mm Petri dishes.
12. 15-ml polystyrene conical centrifuge tubes.
13. 50-ml polystyrene conical centrifuge tubes.
14. 30 G × 1/2 in. needles.
15. Syringes.
16. 0.22-μm filter units.
17. Transfer pipettes: Heat-pulled Pasteur pipettes with opening diameter of 150–200 μm.
18. Micro-dissection glass hooks: Pasteur pipettes are pulled hair thin, and the fine glass needles ends are forged into a hook on fire.

2.3 Reagents

1. High glucose Dulbecco's modified Eagle's medium (DMEM).
2. Fetal bovine serum (FBS).
3. 200 mM L-Glutamine (100×).
4. Antibiotic–antimycotic (100×).
5. Gelatin.
6. Mitomycin C.
7. Phosphate-buffered saline (PBS).
8. 0.05 % Trypsin–EDTA.

9. Knockout™ DMEM.
10. Knockout™ serum replacement (KSR).
11. 100 mM MEM nonessential amino acid solution (NEAA) (100×).
12. β-Mercaptoethanol.
13. bFGF.
14. BIO.
15. Mineral oil.
16. Bovine serum albumin (BSA).
17. Protease.
18. Water.
19. Vector® blue alkaline phosphatase substrate kit III (Vector Laboratories).
20. Paraformaldehyde.
21. Goat serum.
22. Triton X-100.
23. Rabbit anti-Oct-4 polyclonal antibody (Santa Cruz, sc-9081).
24. Rabbit anti-Nanog polyclonal antibody (Novus, ab10626).
25. Mouse anti-SSEA1 monoclonal antibody (DSHB, MC-480).
26. Mouse anti-SSEA4 monoclonal antibody (DSHB, MC-813-70).
27. Alexa Fluor 488 Goat anti-mouse IgG (Molecular Probes, A-11001).
28. Alexa Fluor 488 Goat anti-rabbit IgG (Molecular Probes, A-11008).
29. Alexa Fluor 568 Goat anti-mouse IgG (Molecular Probes, A-11004).
30. Alexa Fluor 568 Goat anti-rabbit IgG (Molecular Probes, A-11011).
31. Hoechst 33342 (Molecular Probes, H1398).
32. Vectashield mounting solution (Vector Laboratories).

2.4 Medium Preparation

1. Bovine embryo culture medium: potassium simplex optimized medium (KSOM) [50] supplemented with 1 % BSA, filter through a 0.22-μm filter unit, aliquot in 10 ml amounts and store at 4 °C for less than 2 weeks (*see* **Note 1**).
2. Bovine embryo transfer medium: HEPES (14 mM)-buffered KSOM (HKSOM) supplemented with 1 % BSA, filter sterilization, aliquot in 10 ml amounts and store at 4 °C for less than 2 weeks.
3. 0.1 % Gelatin: 100 mg gelatin dissolved in 100 ml PBS, autoclaved and store at 4 °C.
4. Mitomycin C: 1 mg/ml dissolved in PBS, filter sterilization, aliquot in 1 ml amounts and store at −20 °C.

5. Protease: 5 mg/ml dissolved in 1 % BSA-HKSOM, filter sterilization, aliquot in 1 ml amounts and store at −20 °C.
6. MEF culture medium: High-glucose DMEM supplemented with 10 % FBS, 1 mM L-glutamine, and 1 % antibiotic–antimycotic, store at 4 °C for less than 2 weeks.
7. Bovine ES cell primary culture medium: 80 % Knockout Dulbecco's modified Eagle's medium supplemented with 20 % knockout serum replacement (KSR), 0.1 mM nonessential amino acid (NEAA), 4 ng/ml bFGF, 5 μM BIO, 0.1 mM β-mercaptoethanol, 1 mM L-glutamine, and 1 % antibiotic-antimycotic, store at 4 °C for less than 2 weeks.

3 Methods

3.1 Preparation of Feeder Layers

1. Thaw one vial of MEF cells onto one 100-mm cell culture dish (*see* **Note 2**), and allow to grow to 80–90 % confluence.
2. Aspirate medium off of the MEFs, and incubate the cells with 5 ml MEF culture medium supplemented 10 μg/ml mitomycin C for 3 h.
3. Add enough 0.1 % gelatin solution to cover the bottom of the cell culture plate for at least 15 min, and remove the solution prior to cell plating.
4. Wash the mitomycin C-treated cells with 5 ml PBS twice, trypsinized and seed onto a 4-well plate with a density of 8×10^4 cells per well. The feeder layer should be ready 1 day before the derivation of bovine primary ES cell colony.

3.2 Preparation of Bovine Mid-stage Blastocysts

1. Both in vivo produced or IVF-derived bovine mid-stage blastocysts can be used for derivation of primary ES cell colony.
2. Fresh in vivo produced embryos collected from Holstein–Friesian cows nonsurgically 7 days after fertilization are transported to the laboratory at 37 °C in 1 % BSA–HKSOM. Some embryos might be morula or early blastocysts, and need to be further cultured to mid-stage blastocysts before derivation.
3. Culture in vivo produced morula and early blastocysts in vitro: wash the embryos thrice in 1 % BSA–KSOM and then culture at 38.6 °C in humidified atmosphere of 5.6 % CO_2 until mid-stage blastocysts (*see* **Note 3**).

3.3 Derivation of Primary Bovine ES Colony

1. Change the MEF culture medium with 750 μl ES cell culture medium 2 h before the treatment of blastocysts.
2. Remove the zona pellucida of the blastocyst by incubation in 5 mg/ml protease for a few minutes (*see* **Note 4**), followed by gentle pipetting and washing with HKSOM thrice (Fig. 1).
3. Transfer one denuded blastocyst into one well of 4-well tissue culture plate, and culture in a humidified atmosphere of 5.6 %

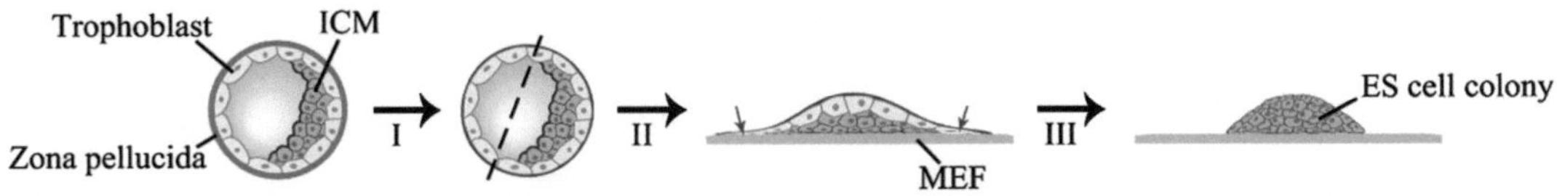

Fig. 1 Diagram for the derivation of bovine primary ES cell colony. I: The zona pellucida of mid-stage bovine blastocyst is removed. II: The blastocyst is incised into two parts: the part with trophectoderm (TE) cells that is discarded and the part with entire ICM. The ICM part is elaborately spread and fixed onto the MEF cells by gently pressing the residual TE cells around the ICM using a 32-gauge needle. III: The primary bovine ES cell colony, ICM outgrowth, appears within 2–7 days

CO_2 in air at 38.6 °C for 1 h for recovery. The cavity of some blastocysts might be broken, and these embryos need longer time, usually 3–12 h, to expand again (*see* **Note 5**).

4. Incise the blastocyst into two parts in the original culture well: the part with entire ICM and the part with trophectoderm (TE) cells that will be removed from the well. The ICM part is elaborately spread and fixed onto feeder cells by gently riveting the residual TE cells around the ICM using a 32-gauge needle (*see* **Note 6**). Consequently, the ICM is "seeded" on the feeder cells (Fig. 1).
5. Leave the culture untouched in a humidified atmosphere of 5.6 % CO_2 in air at 38.6 °C for 2 days. Then, change the culture medium every 2 days.
6. The dome-shape ICM-derived outgrowths with well-defined boundary can be observed under a stereomicroscope within 2–7 days. The cells in the primary colony exhibit clear large nuclei with one or more prominent nucleoli, and are tightly packed within the multiple layers, characteristic of mouse ES cells (Fig. 2a, b). In our lab, 87 primary bovine ES cell colonies were isolated from 110 blastocysts (79.1 % success rate) [49]. Moreover, this "separate and seed" method has been successfully used to isolate primary ovine ES cell colonies [51].

3.4 Passage of Bovine ES Cells

1. Prepare the MEF feeder cells in 4-well plate as described above 1 day before, and change the MEF culture medium with 750 μl ES culture medium 2 h before passage.
2. Ideal colonies are micro-dissected into clumps of about 100 cells using the micro-dissection glass hooks. The cell clumps are transferred to the 4-well plate using glass pipettes by 1:4 ratio split, and designate as passage 1 (P1). The cell clumps from one blastocyst are plated into the same well, and number of the newly formed colony is recorded.
3. Growth medium is changed every 2 days.
4. Mechanically passage the bovine ES cell colonies every 5–7 days when the colonies reach ~90 % confluence. In our case,

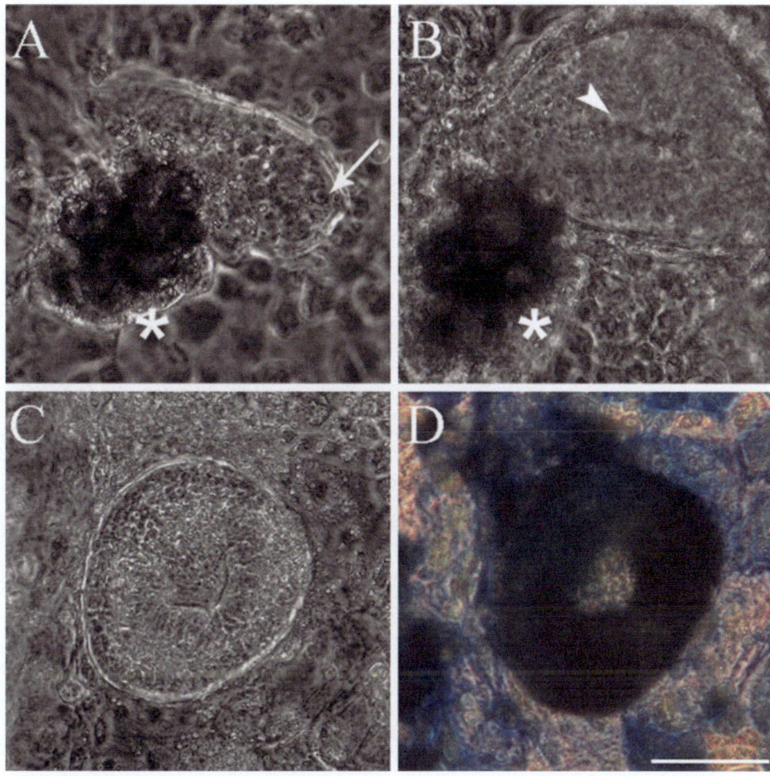

Fig. 2 Morphology and AP activity of bovine primary ES cell colony. (**A**, **B**) Morphology of bovine primary ES colony cultured for 2 days (**A**) and 5 days (**B**). The colony at day 2 displays morphology typical of ES cells with prominent nucleoli (indicated by *arrow*), high ratio of nucleus to cytoplasm and well-defined colony boundary distinguishable from MEF feeder cells, and the colony forms dome-like structure (indicated by *arrowhead*) at day 5. The TE cells are pushed aside forming a dark clump (indicated by *asterisk*). (**C**, **D**) Primary ES colony (**C**) demonstrates strong AP activity (**D**). Some TE cells also show AP activity. Scale bar = 100 μm. Reproduced from Cao et al. [49] with permission from John Wiley and Sons

the bovine ES cell colony can self-renew to passage 10. The colony at passage 0–4 is small and displays clear boundary with feeder cells similar to human ES colony (Fig. 3A, B). With passaging, the peripheral cells of bovine ES cell colony begin to differentiate, and the boundary becomes undistinguishable gradually (Fig. 3C, D).

3.5 In Vitro Characterization of Bovine ES Cells

3.5.1 Alkaline Phosphatase (AP) Activity Assay

1. The activity of AP is detected using the Vector® Blue Alkaline Phosphatase Substrate Kit III from Vector Laboratories, following the manufacture's protocol.
2. The positive signals of AP activity are captured with a color digital camera. The primary bovine ES cell colony displays high AP activity in the ICM outgrowth and some AP activity in surrounding TE cells (Fig. 2C, D).

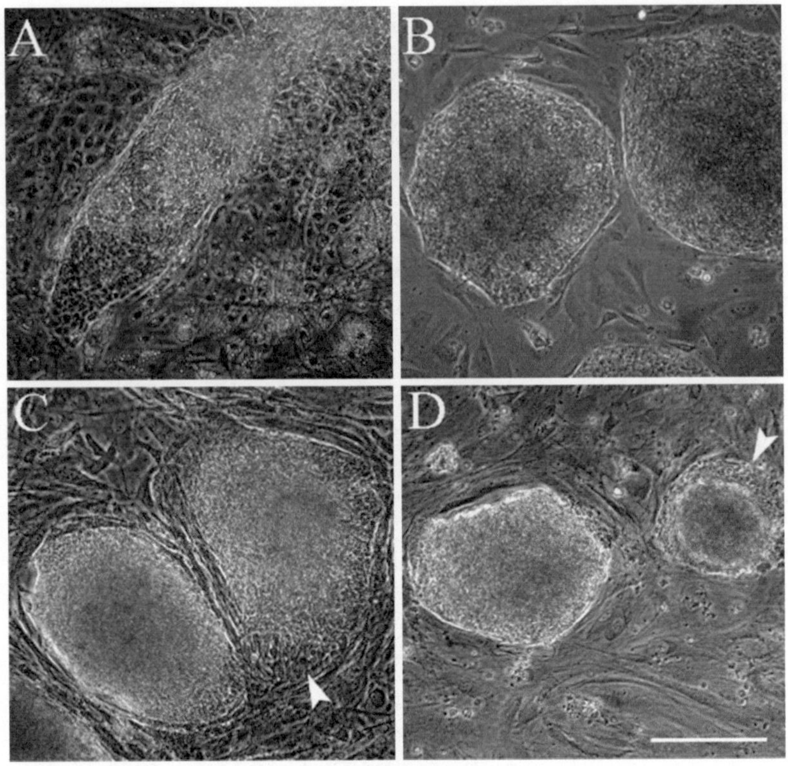

Fig. 3 Morphology of bovine ES cell colonies at P0 (**A**), P4 (**B**), P6 (**C**), and P10 (**D**) cultured in the medium supplemented with 4 ng/ml bFGF and 5 μM BIO. Bovine ES cell colonies maintain undifferentiated state at early passages (**A**, **B**), but exhibit some differentiation at the edge of the colonies by passage 6 and 10 (**C**, **D**, indicated by *arrowheads*). Scale bar = 200 μm

3.5.2 Immunofluorescence Staining of ES Cell Pluripotency Markers

1. Fixation: The bovine ES cells are washed with 500 μl PBS twice and fixed in 500 μl freshly prepared paraformaldehyde (3.7 % w/v in PBS) at 4 °C for 15 min.
2. Permeabilization: The cells are rinsed with 500 μl blocking buffer (3 % goat serum in PBS) thrice for 15 min each, and permeabilized with 500 μl 0.1 % Triton X-100 in blocking buffer for 15 min.
3. Blocking: The cells are rinsed with 500 μl blocking buffer thrice for 15 min and incubated in 500 μl blocking solution for 1 h.
4. First antibody: The cells are probed with 100 μl diluted primary antibody (1:200 dilution in blocking solution) at 4 °C overnight.
5. Second antibody: The cells are washed with 500 μl blocking solution thrice for 15 min each, and incubated with 100 μl diluted appropriate secondary antibodies conjugated to Alexa

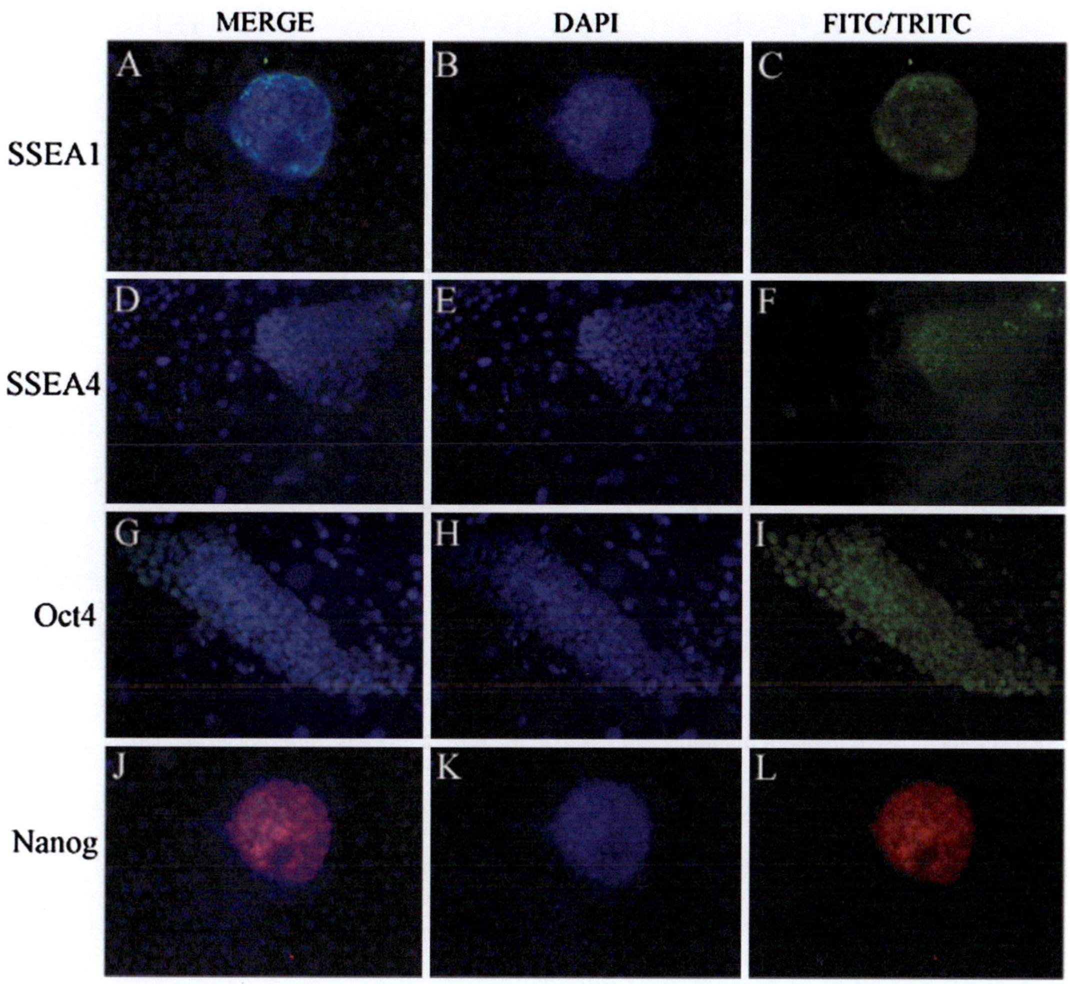

Fig. 4 Representative immunofluorescence images showing the expression of SSEA1 (**A–C**), SSEA4 (**D–F**), Oct4 (**G–I**), and Nanog (**J–L**) in the primary bovine ES colony. SSEA1 (FITC, *green*) and SSEA4 (FITC, green) are expressed on the cell surface, and Oct4 (FITC, *green*) and Nanog (TRITC, *red*) are localized to the nuclei. The TE cells around the primary ES colonies and MEF feeder cells served as negative controls do not express the pluripotent markers. *Blue*, nuclei stained by DAPI. Reproduced from Cao et al. [49] with permission from John Wiley and Sons

Fluor 488 or 568 (1:100 dilution in blocking solution) for 1 h at room temperature, followed by washing again.

6. Counterstaining: The cells are counterstained with 0.2 μg/ml Hoechst 33342 in 100 μl Vectashield mounting medium.
7. Picture capture: Immunofluorescence of Oct4, Nanog, SSEA1, and SSEA4 are digitally imaged with appropriate filters under a Leica fluorescence microscope. The primary bovine ES cells specifically express SSEA1 (Fig. 4A–C) and SSEA4 (Fig. 4D F) on the cell surface, and Oct4 (Fig. 4G–I) and Nanog (Fig. 4J–L) in the nuclei, whereas the TE cells surrounding the primary

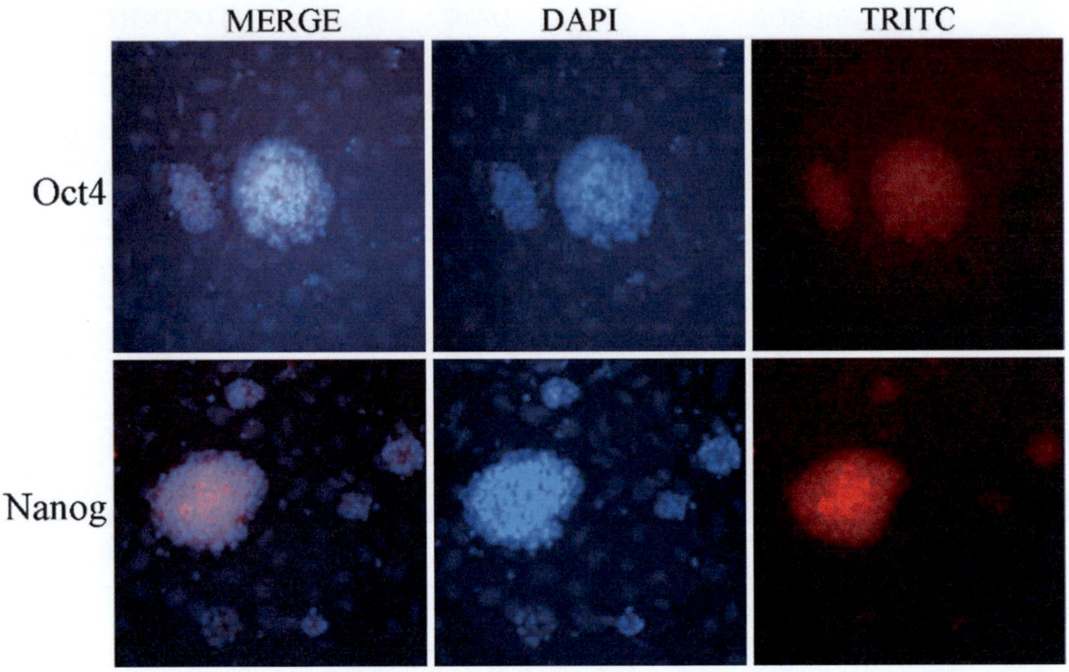

Fig. 5 Expression of pluripotent genes Oct4 (**a**–**c**) and Nanog (**d**–**f**) in the bovine ES colony at P4. Oct4 (TRITC, *red*) and Nanog (TRITC, *red*) are markedly expressed in the bovine ES cells. *Blue*, nuclei stained by DAPI

culture and the mouse embryonic fibroblasts (MEF) cells do not express these markers (Fig. 4). Furthermore, the bovine ES colonies at passage 4 express Oct4 and Nanog as shown by immunostaining (Fig. 5).

4 Notes

1. All the culture medium should be stored at 4 °C and used within 2 weeks. The medium beyond 2 weeks might reduce the blastocyst rate, affect the primary bovine ES cell colony formation, and induce the ES cell differentiation.
2. It is better to use early passage MEF cells (P1–P3) for the isolation and culture of bovine ES cells. According to our unpublished data, early passages of MEF cells (P1–P3) cultured at 37 or 38.6 °C display no obvious difference in morphology and expression levels of some apoptotic genes such as Bcl2, Bax, and Hsp70. In addition, compared with bovine embryonic fibroblasts (BEF), MEF can better support the growth of bovine ES cell colony.
3. It usually takes 12–24 h for the morula to reach mid-stage blastocyst and 2–6 h for early blastocyst.

4. The compositions of zona pellucida of in vivo produced and IVF-derived bovine mid-stage blastocysts seem different. It usually takes 1–2 min to completely digest the zona pellucida of IVF-derived bovine embryos and 3–5 min to the in vivo produced embryos.
5. It is recommended to culture the blastocysts to fully expand again. If not, it is difficult to distinguish the ICM from TE cells. Even you can distinguish the ICM; it is hard to incise the ICM part from the TE cells.
6. Press the needle very gently! If too hard, the needle would break the TE cells and feeder cells, so that TE cells cannot attach to the feeder cells.

Acknowledgements

This study was supported by Ministry of Science and Technology China (863 plan 2010AA10A103), the National Basic Research Program of China (2009CB941000 and 2011CBA01002) and China National Natural Science Foundation.

References

1. Evans MJ, Kaufman MH (1981) Establishment in culture of pluripotential cells from mouse embryos. Nature 292:154–156
2. Martin GR (1981) Isolation of a pluripotent cell line from early mouse embryos cultured in medium conditioned by teratocarcinoma stem cells. Proc Natl Acad Sci U S A 78:7634–7638
3. Smith AG (2001) Embryo-derived stem cells: of mice and men. Annu Rev Cell Dev Biol 17:435–462
4. Heins N, Englund MC, Sjoblom C, Dahl U, Tonning A, Bergh C, Lindahl A, Hanson C, Semb H (2004) Derivation, characterization, and differentiation of human embryonic stem cells. Stem Cells 22:367–376
5. Lagasse E, Shizuru JA, Uchida N, Tsukamoto A, Weissman IL (2001) Toward regenerative medicine. Immunity 14:425–436
6. Keefer CL, Pant D, Blomberg L, Talbot NC (2007) Challenges and prospects for the establishment of embryonic stem cell lines of domesticated ungulates. Anim Reprod Sci 98:147–168
7. Talbot NC, Blomberg LA (2008) The pursuit of ES cell lines of domesticated ungulates. Stem Cell Rev 4:235–254
8. Nowak-Imialek M, Kues W, Carnwath JW, Niemann H (2011) Pluripotent stem cells and reprogrammed cells in farm animals. Microsc Microanal 17:474–497
9. Stice SL, Strelchenko NS, Keefer CL, Matthews L (1996) Pluripotent bovine embryonic cell lines direct embryonic development following nuclear transfer. Biol Reprod 54:100–110
10. Strelchenko NS (1996) Bovine pluripotnent stem cells. Theriogenology 45:131–140
11. Cibelli JB, Stice SL, Golueke PJ, Kane JJ, Jerry J, Blackwell C, Ponce de Leon FA, Robl JM (1998) Transgenic bovine chimeric offspring produced from somatic cell-derived stem-like cells. Nat Biotechnol 16:642–646
12. Iwasaki S, Campbell KH, Galli C, Akiyama K (2000) Production of live calves derived from embryonic stem-like cells aggregated with tetraploid embryos. Biol Reprod 62:470–475
13. Kitiyanant Y, Guocheng JS, Pavasuthipaisit K (2000) Establishment and long-term maintenance of bovine embryonic stem cell lines using mouse and bovine mixed feeder cells and their survival after cryopreservation. Sci Asia 26:81–86
14. Mitalipova M, Beyhan Z, First NL (2001) Pluripotency of bovine embryonic cell line derived from precompacting embryos. Cloning 3:59–67
15. Saito S, Sawai K, Ugai H et al (2003) Generation of cloned calves and transgenic chimeric embryos from bovine embryonic stem-like cells. Biochem Biophys Res Commun 309:104–113

16. Wang L, Duan E, Sung LY et al (2005) Generation and characterization of pluripotent stem cells from cloned bovine embryos. Biol Reprod 73:149–155
17. Leal de Freitas ER, Sanches BV, Gambarini ML et al (2011) Embryonic stem-like cells derived from in vitro produced bovine blastocysts. Braz Arch Biol Technol 54:495–502
18. Lim ML, Vassiliev I, Richings NM et al (2011) A novel, efficient method to derive bovine and mouse embryonic stem cells with in vivo differentiation potential by treatment with 5-azacytidine. Theriogenology 76:133–142
19. Jin M, Wu A, Dorzhin S et al (2012) Culture conditions for bovine embryonic stem cell-like cells isolated from blastocysts after external fertilization. Cytotechnology 64:379–389
20. Takahashi K, Yamanaka S (2006) Induction of pluripotent stem cells from mouse embryonic and adult fibroblast cultures by defined factors. Cell 126:663–676
21. Takahashi K, Tanabe K, Ohnuki M et al (2007) Induction of pluripotent stem cells from adult human fibroblasts by defined factors. Cell 131:861–872
22. Yu J, Vodyanik MA, Smuga-Otto K et al (2007) Induced pluripotent stem cell lines derived from human somatic cells. Science 318:1917–1920
23. Liu H, Zhu F, Yong J, Zhang P et al (2008) Generation of induced pluripotent stem cells from adult rhesus monkey fibroblasts. Cell Stem Cell 3:587–590
24. Liao J, Cui C, Chen S et al (2009) Generation of induced pluripotent stem cell lines from adult rat cells. Cell Stem Cell 4:11–15
25. West FD, Terlouw SL, Kwon DJ et al (2010) Porcine induced pluripotent stem cells produce chimeric offspring. Stem Cells Dev 19:1211–1220
26. Ren J, Pak Y, He L et al (2011) Generation of hircine-induced pluripotent stem cells by somatic cell reprogramming. Cell Res 21:849–853
27. Sartori C, DiDomenico AI, Thomson AJ et al (2012) Ovine-induced pluripotent stem cells can contribute to chimeric lambs. Cell Reprogram 14:8–19
28. Huang B, Li T, Alonso-Gonzalez L, Gorre R et al (2011) A virus-free poly-promoter vector induces pluripotency in quiescent bovine cells under chemically defined conditions of dual kinase inhibition. PLoS One 6:e24501
29. Han X, Han J, Ding F et al (2011) Generation of induced pluripotent stem cells from bovine embryonic fibroblast cells. Cell Res 21:1509–1512
30. Sumer H, Liu J, Malaver-Ortega LF et al (2011) NANOG is a key factor for induction of pluripotency in bovine adult fibroblasts. J Anim Sci 89:2708–2716
31. Davis S, Aldrich TH, Stahl N et al (1993) LIFR beta and gp130 as heterodimerizing signal transducers of the tripartite CNTF receptor. Science 260:1805–1808
32. Yoshida K, Chambers I, Nichols J et al (1994) Maintenance of the pluripotential phenotype of embryonic stem cells through direct activation of gp130 signalling pathways. Mech Dev 45:163–171
33. Niwa H, Burdon T, Chambers I, Smith A (1998) Self-renewal of pluripotent embryonic stem cells is mediated via activation of STAT3. Genes Dev 12:2048–2060
34. Ying QL, Nichols J, Chambers I, Smith A (2003) BMP induction of Id proteins suppresses differentiation and sustains embryonic stem cell self-renewal in collaboration with STAT3. Cell 115(3):281–292
35. Qi X, Li TG, Hao J et al (2004) BMP4 supports self-renewal of embryonic stem cells by inhibiting mitogen-activated protein kinase pathways. Proc Natl Acad Sci U S A 101:6027–6032
36. Vallier L, Alexander M, Pedersen RA (2005) Activin/Nodal and FGF pathways cooperate to maintain pluripotency of human embryonic stem cells. J Cell Sci 118:4495–4509
37. Brown S, Teo A, Pauklin S et al (2011) Activin/Nodal signaling controls divergent transcriptional networks in human embryonic stem cells and in endoderm progenitors. Stem Cells 29:1176–1185
38. Okita K, Yamanaka S (2006) Intracellular signaling pathways regulating pluripotency of embryonic stem cells. Curr Stem Cell Res Ther 1:103–111
39. Miki T, Yasuda SY, Kahn M (2011) Wnt/beta-catenin signaling in embryonic stem cell self-renewal and somatic cell reprogramming. Stem Cell Rev 7:836–846
40. Sato N, Meijer L, Skaltsounis L et al (2004) Maintenance of pluripotency in human and mouse embryonic stem cells through activation of Wnt signaling by a pharmacological GSK-3-specific inhibitor. Nat Med 10:55–63
41. Besser D (2004) Expression of nodal, lefty-a, and lefty-B in undifferentiated human embryonic stem cells requires activation of Smad2/3. J Biol Chem 279:45076–45084
42. Ying QL, Wray J, Nichols J (2008) The ground state of embryonic stem cell self-renewal. Nature 453:519–523

43. Niwa H, Ogawa K, Shimosato D, Adachi K (2009) A parallel circuit of LIF signalling pathways maintains pluripotency of mouse ES cells. Nature 460:118–122
44. Sudheer S, Bhushan R, Fauler B et al (2012) FGF inhibition directs BMP4-mediated differentiation of human embryonic stem cells to syncytiotrophoblast. Stem Cells Dev 21(16): 2987–3000
45. Cheng J, Dutra A, Takesono A, Garrett-Beal L, Schwartzberg PL (2004) Improved generation of C57BL/6J mouse embryonic stem cells in a defined serum-free media. Genesis 39: 100–104
46. Nagafuchi S, Katsuta H, Kogawa K et al (1999) Establishment of an embryonic stem (ES) cell line derived from a non-obese diabetic (NOD) mouse: in vivo differentiation into lymphocytes and potential for germ line transmission. FEBS Lett 455:101–104
47. Reubinoff BE, Pera MF, Fong CY et al (2000) Embryonic stem cell lines from human blastocysts: somatic differentiation in vitro. Nat Biotechnol 18:399–404
48. Thomson JA, Itskovitz-Eldor J, Shapiro SS, Waknitz MA, Swiergiel JJ, Marshall VS, Jones JM (1998) Embryonic stem cell lines derived from human blastocysts. Science 282: 1145–1147
49. Cao S, Wang F, Chen Z et al (2009) Isolation and culture of primary bovine embryonic stem cell colonies by a novel method. J Exp Zool A Ecol Genet Physiol 311(5):368–376
50. Liu L, Keefe DL (2007) Nuclear transfer methods to study aging. Methods Mol Biol 371:191–207
51. Zhao Y, Lin J, Wang L et al (2011) Derivation and characterization of ovine embryonic stem-like cell lines in semi-defined medium without feeder cells. J Exp Zool A Ecol Genet Physiol 315:639–648

Chapter 10

Analysis of Molecular Markers for Staging Peri-gastrulating Bovine Embryos

Isabelle Hue, Severine A. Degrelle, and Christoph Viebahn

Abstract

Whole-mount in situ hybridization (WISH) is a method to visualize gene expression through hybridization of in vitro synthesized riboprobes to cellular mRNAs. WISH has been used in developmental biology for decades and was adapted to many species, especially for model organisms of developmental biology. The method has evolved, from analyzing small embryo batches to large-scale gene expression screenings, using: (1) manual or automated protocols, (2) colorimetric or fluorescent detection of the hybridized riboprobes, and (3) individual or systemic image acquisition and storage. As for bovine embryo staging, the in situ hybridization of whole embryonic discs has both proved useful and efficient, provided that a few improvements were brought to the in vitro riboprobe synthesis.

Key words WISH, In situ hybridization, Embryo or tissue preparation, In vitro riboprobe synthesis, Digoxigenin-RNA labelled probe, Colorimetric detection

1 Introduction

Embryonic staging often relies on morphological traits [1, 2] molecular patterns or both. In the latter cases, molecular patterns were obtained using whole-mount in situ hybridization (WISH) with digoxigenin-RNA labelled probes [3, 4], based on temporally and spatially restricted gene expression profiles (cow [5], rabbit [6, 7], sheep [8], pig [9]). For this purpose, hybridization of riboprobes to cellular mRNAs is classically detected by colorimetric reactions using alkaline phosphatase (AP) or horseradish peroxidase (HP) and their substrates (NBT/BCIP or BM Purple). WISH with fluorescent probes was also developed, as well as simultaneous multiple mRNA detections (chick [10], zebrafish [11]). Here, we describe the WISH protocol as it was used in our recent work on bovine embryos using a manual protocol, simplified riboprobe synthesis, and colorimetric detection [12, 19]. However, we are well aware that WISH has been used for large-scale gene expression screenings in model organisms of developmental biology

Ramiro Alberio (ed.), *Epiblast Stem Cells: Methods and Protocols*, Methods in Molecular Biology, vol. 1074,
DOI 10.1007/978-1-62703-628-3_10, © Springer Science+Business Media New York 2013

(xenopus [13], drosophila [14]); also, in situ hybridization on serial tissue sections has emerged to maximize the number of probes per mammalian embryo (mouse [15], cow [16]). Whatever the objectives, the basics of WISH are the same, i.e., detecting cellular mRNA by in situ hybridization with antisense labelled riboprobes applying four main consecutive steps: embryo or tissue preparation, RNA probe synthesis, in situ hybridization, and labelled riboprobe detection.

2 Materials

2.1 Embryo Preparation

Once embryos are harvested using a modified IMV catheter and a gentle uterus flush with warm PBS [5, 16, 19], rinse them in fresh PBS before treatment. Then:

1. Fix the embryos in a solution 4 % paraformaldehyde in PBS for 2–3 h on ice.
2. Wash three times with PBS for 5 min each on ice.
3. Dehydrate embryos with graded series of cold methanol solutions in PBS for 5 min each: 25, 50, 75, 100 % (on ice).
4. Rinse once more in methanol 100 % and store the embryos in 100 % methanol at −20 °C until needed in plastic tubes or classical freezing vials.

2.2 Synthesis of RNA Probes

2.2.1 cDNA Template

This protocol describes an easy alternative to prepare a cDNA template for in vitro transcription by using PCR on a plasmid or a bacterial culture. This was initially developed for large-scale probe synthesis [17] and is even quicker now since PCR products can easily be purified on specific columns prior to riboprobe synthesis.

1. Select the gene of interest from a cDNA library or clone it in a plasmid (Fig. 1a).
2. Amplify the cDNA fragment with the appropriate primers and annealing temperatures according to the plasmid used (Fig. 1b, Table 1).
3. Purify the amplified product with a QIAquick PCR Purification column (Qiagen) and check 50–100 ng of it on a 1–2 % agarose gel (a single band of the expected size is expected on denaturing RNA gels).

2.2.2 In Vitro Transcription

To define the sense and antisense probes, check the orientation of the cloned cDNA by sequencing or by verifying the cloning strategy of the cDNA library (Fig. 1a, b). The antisense probe is the one that hybridizes to the mRNA within the cell, whereas the sense probe is the one that cannot hybridize to it and is thus the negative control of the ISH.

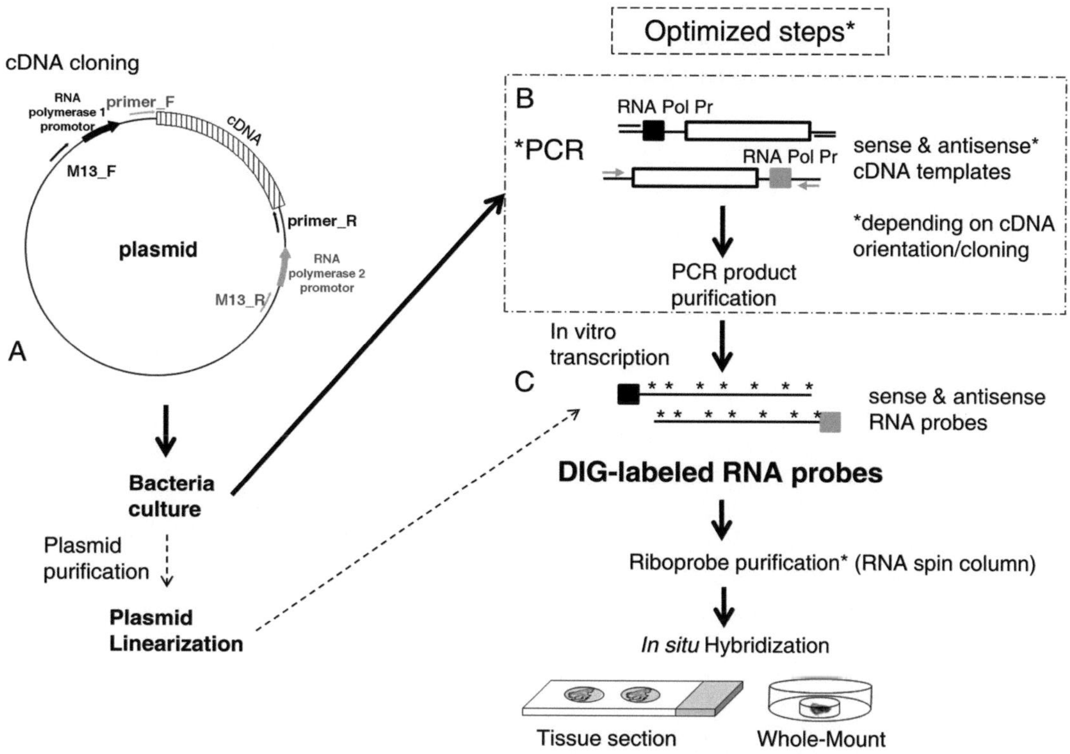

Fig. 1 Simplified riboprobe synthesis. The modifications as compared to Weisheit et al. [4] include (1) a PCR amplification of the cDNA template—starting from a purified plasmid or a bacterial culture (*A*) using sequences from the plasmid backbones, on each side of the cDNA, as illustrated in (*B*) and (2) two purification steps with spin columns, one for the amplified cDNA template and one for the labelled RNA probes. RNA probe synthesis (*C*) was performed as described [3–9]. Table 1 provides a set of primer pairs for the plasmids we have been using so far

1. Use an in vitro transcription kit to perform the riboprobe synthesis (Fig. 1c) as follows:
   - Amplified cDNA template (100–200 ng).
   - 10× Transcription buffer.
   - 10× DIG-labelling mix.
   - RNA polymerase.
   - RNase-free H_2O up to 20 μl.
2. Incubate at 37 °C for 2 h.
3. Add RNA polymerase for another incubation at 37 °C for 1 h, to maximize the yield of the riboprobe synthesis.
4. Test an aliquot of the reaction on a 1–2 % agarose gel, to check the success of the synthesis: one band at the right size with a denaturing RNA gel (Northern blot buffer); one fuzzy band or even several bands with a non-denaturing RNA gel (TBE buffer).

Table 1
Set of DNA primer pairs, derived from plasmid backbones, to amplify cDNA templates according to their cloning orientation, prior to labelled RNA probe synthesis using an in vitro transcription kit

	To be transcribed with RNA polymerase	DNA template synthesis with primers	Referenced in previous work
pGEM-T easy	T7 Sp6	M13_F+primer1_R M13_R+primer1_F	[12]
pGEM-T	T7 Sp6	M13_F+primer2_R M13_R+primer2_F	Unpublished
pGEMZf11(+)	T7 Sp6	M13_F+primer3R M13_R+primer3_F	[19]
pBluescript SK(+/−) et II SK(+/−)	T7 T3	M13_F+primer4_R M13_R+primer4_F	Unpublished
pT3T7Pac	T7 T3	M13_R+primer5_F M13_F+primer5_R	[19]
Primers	5′→3′ sequence	Tm (°C)	
M13F	GTA AAA CGA CGG CCA GTG A	56.7	
M13R	CAC AGG AAA CAG CTA TGA CC	57.3	
primer1_F	CGG CCG CGG GAA TTC GAT	60.5	
primer1_R	GGC CGC GAA TTC ACT AGT GAT	59.8	
primer2_F	CGC CAT GGC CGC GGG ATT	62.8	
primer2_R	AGG CGG CCG CAC TAG TGA	60.5	
primer3_F	ATT GGC CAA GTC GGC CGA	58.2	
primer3_R	CTC AAG CTT ATG CAT GCG G	56.7	
primer4_F	TGC TTG CGG CCG CAT TTG TTT	59.8	
primer4_R	AAA AGC TGG AG CTC CAC CG	58.8	
primer5_F	GAA TAA GCT TGC GGC CGC	59.8	
primer5_R	GCC CTC GAG GCC AAG AAT	58.2	

5. Purify the riboprobe on a mini Quick Spin RNA Column (Roche) and add 1 μl RNAse inhibitor (Ambion) before storage at −20 °C (−80 °C recommended for long-term storage).

2.3 Buffers, Reagents

Water must be deionized, autoclaved and RNAse-free DNAse-free. Similarly, all containers must be clean and free of ribonucleases.

Referring to the protocol (*see* Subheading 3.1), calculate the adequate volume of each buffer, depending on the experimental design, number of probes and number of washes, based on a volume of 1.2 ml per well. Prepare them based on the appropriate

Table 2
Stock solutions to be used in the WISH protocol for the WISH solutions

Stock solutions	Storage
PFA 4 %, in PBS	4 °C
Tween 20, 10 % in PBS	4 °C
Proteinase K, 10 mg/ml in RNAse-free dH_2O	-20 °C
Glutaraldehyde 50 %	Room temp.
Glycine, 100 mg/ml	-20 °C
20× SSC, pH 7	Room temp.
SDS 10 %	Room temp.
Yeast tRNA, 10 mg/ml	-20 °C
MAB: Maleic acid 100 mM–NaCl 150 mM, pH 7.5	Room temp.
NTMT: Tris 100 mM–NaCl 100 mM–$MgCl_2$ 50 mM–Tween 20 0.1 %	Room temp.
Heparin: 10 mg/ml	-20 °C
BBR: 20 mg/ml in MABT	-20 °C

stock solutions (Table 2) as well as detailed recipes, as previously reported [3, 17].

- PBT 1×: Tween 20, 0.1 % in PBS.
- Methanol series: 75, 50, 25 %, volume to volume in PBT 1×.
- Proteinase K, 10 μg/ml, in PBT 1×.
- Post-fixation solution: PFA 1 %–Tween 20 0.1 %–Glutaraldehyde 0.2 %, in PBS.
- Post-fixation washes: Glycine 2 mg/ml, in PBT 1×.
- Solution I: Formamide 50 %, SSC 5× pH 7, SDS 1 % in RNAse-free dH_2O.
- Solution III: Formamide 50 %, SSC 2× pH 7 in RNAse-free dH_2O.
- Pre-hybridization solution: yeast tRNA 100/ml, heparin 50 μg/ml in RNAse-free dH_2O.
- Hybridization solution: riboprobe at 1 μg/ml in the pre-hybridization solution.
- Post-hybridization washes: MABT 1×. Dilute Tween 20 10 % at 0.1 % in MAB.
- Blocking solution: sheep serum 15 % and BBR 2 %, in MABT 1×.

2.4 Equipment

- Hybridization oven (65–70 °C).
- Rocking platform (all days).
- Fume hood (methanol, post-fixation solution, solution I, solution II, pre-hybridization, and hybridization solutions).
- Tubes with fine mesh net [4] or 12-well plates with appropriate buckets with fine mesh net [17].
- PCR machine (riboprobe denaturation, 95 °C).
- Stereomicroscope, Camera (color development and evaluation of the data).

3 Methods

In all the steps preceding in situ hybridization, the degradation of cellular RNA has to be prevented to ensure good results. Therefore, gloves must be worn at all steps and containers must be clean, especially when reused (12-well plates and buckets). Then, treat them as detailed by Shimizu et al. [17].

3.1 Protocol

First of all, transfer the embryos from the vials where they were stored (in 100 % methanol) to the 12-well plates (in 100 % methanol), according to the ISH design. To do so, use a soft plastic pipette or a 200–1,000 μl tip that was cut to make a wide opening.

Day 1: Pre-hybridization and hybridization

1. Rehydrate the embryos using a decreasing methanol–PBT series: 75, 50, 25, 0 % for 5 min each, on ice, and wash two times with PBT, 5 min each, on ice.
2. Digest with 10 μg/ml of proteinase K in PBT, at room temperature, according to the stage of the embryo: 5 min for the youngest ones, 10 min for the oldest ones (somites stages and older).
3. Wash two times in PBT, at room temperature.
4. Postfix in 1 % PFA–Tween 0.1–0.2 % glutaraldehyde, in PBS for 20 min at room temperature. Due to carcinogenic compounds, perform **steps 4–16** in a fume hood.
5. Set the oven at 70 °C and warm the hydridization solution at 70 °C* (*70 °C for homologous probes and embryo species; 65 °C for heterologous probes and species) (Fig. 2).
6. Wash once in 2 mg/ml Glycine in PBT and rinse two times in PBT (for 5 min at room temperature).
7. Wash once in 50 % PBT–50 % solution I.

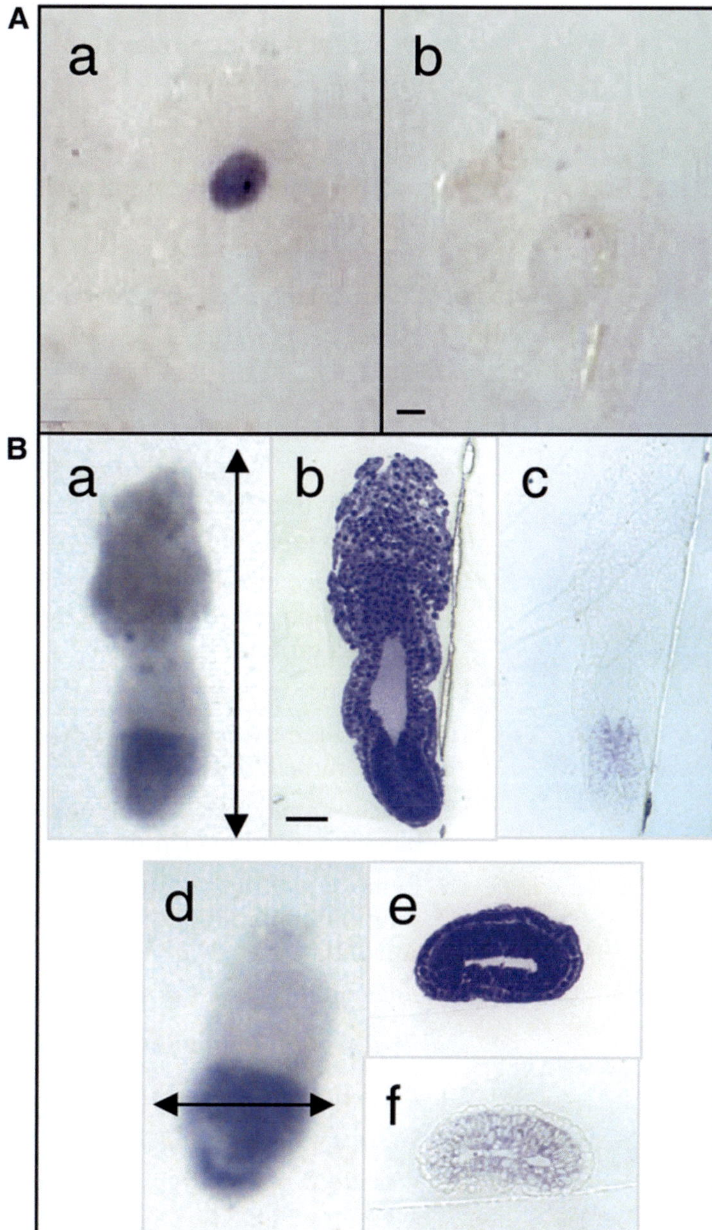

Fig. 2 Heterologous WISH with an Oct-4 probe. (**a**) Embryonic specificity of the bovine oct-4 probe on bovine day-14 conceptuses: (*a*) DIG-labelled antisense probe, (*b*) DIG-labelled sense probe. (**b**) Embryonic specificity of the bovine Oct-4 probe on day 6.5 mouse embryos. Only epiblast cells are labelled (*a*, *d*) as confirmed by sagittal (*b*, *c*) and transverse (*e*, *f*) Technovit® sections. *Black arrows* indicate the position and plane of section of sagittal and transverse sections shown. Magnification bars: 100 μm

8. Pre-hybridize the embryos at 70 °C by incubation in pre-hybridization Solution I for 1 h or longer.
9. Denature the riboprobes at 95 °C for 5 min, in a final volume of 10 µl (DIG-labelled RNA + Rnase-free H_2O). Chill on ice as soon as possible and add to the Hybridization mix after a quick spin.
10. Incubate samples overnight at 70 °C in 1.2 ml of Hybridization Mix containing 1 µg/ml of denatured riboprobe per well, in a wet chamber (50 % water–50 % formamide).
11. Warm Solution I at 70 °C overnight for the first washes on day 2.

Day 2: Post-hybridization washes and preabsorption with the anti-DIG antibody

12. Rinse twice, quickly (1 min) in Solution I at 70 °C.
13. Wash twice in solution I at 70 °C, for 30 min each. While washing, prepare solution III and warm it at 65 °C (*60 °C in heterologous conditions).
14. Wash twice in solution III at 65 °C for a minimum of 30 min each, or longer, to get rid of the free probe and thus minimize the background.
15. Wash once in 50 % solution III–50 % MABT.
16. Wash three times in MABT (the last one can be done outside of the fume hood).
17. Block samples for 1 h or more at room temperature in the blocking solution (1.6 ml/well instead of 1.2 ml).
18. Remove half this volume and replace it by 0.8 ml of blocking solution containing 1/2,000 horseradish peroxidase coupled anti-DIG antibody.
19. Incubate overnight at 4 °C (±16 h).

Day 3: Post-antibody washes and color development

20. Wash three times in MABT, 1 min each.
21. Wash seven times in MABT for 30 min each.
22. Wash four times, 10 min each, in NTMT and leave BM purple at room temperature.
23. Place the embryos in BM purple and incubate in the dark, at room temperature, for 30 min to 12 h. Check the staining regularly under the stereomicroscope as it can come up quickly, but do not expose the samples to light for prolonged periods of time.
24. To stop the reaction, wash and store in NTMT or resume reaction by restaining with BM purple.
25. Wash several times in NTMT and at last, in PBS.
26. Take photographs and store the samples in glycerol before embedding and sectioning to precisely locate the labelling within the embryo (details in ref. 18 and Fig. 2).

3.2 Troubleshooting

Some of the most frequent problems in WISH are mentioned here; they occur either during riboprobe synthesis (lack of DNA template or riboprobe) or during in situ hybridization (no signal, too much background or positive sense probe). Below are a few guidelines and points for optimization:

3.2.1 Probe Size, Specificity, Concentration

The size of the probe is an important parameter since short probes (less than 200 bp) carry fewer label, and thus are less easily detected when the level of cellular transcript is low. In this case, the following two options may be of help: (1) design longer probes and (2) use a combination of 2–3 probes of similar size. So far, we used both these options but longer probes often give better results on tissue sections than in WISH due to problems in the balance between sample permeabilization and riboprobe penetration when probes exceed 1 kb.

Even though WISH with heterologous riboprobes does work (Fig. 2), this varies between genes and depends largely on the cross-species conservation of the sequence, the 3′ untranslated regions (UTRs) being mostly species specific and most variable. The highest WISH specificity is thus guaranteed by the choice of homologous probes, which in turn allow highest hybridization temperature (70 °C) and most stringent post-hybridization washes.

Depending on the level of expression of the cellular transcript to be visualized, reducing or increasing the "basic" probe concentration (1 μg/ml of hybridization mix) is also an option when the use of this concentration gives strong or faint signals.

3.2.2 Probe Purification

As for any hybridization technique (Northern blot, Southern blot), the better purified the probe is, the lower is the background and the "sharper" the signal. When the background generated by an antisense probe is high, use a longer Spin RNA column (Roche) or increase the difference in size between the RNA probe and the free labelled UTPs. As mentioned above regarding the size of the probe, a riboprobe of less than 100 bp may also cause problems.

3.2.3 Length of the Hybridization, Post-hybridization Washes, Color Development

When the background generated by an antisense probe is high, the other options are to: (1) improve post-hybridization washes, increase their numbers, lengths, or both, (2) carefully monitor color development, or (3) reduce hybridization time. As an example, for highly abundant transcripts, we successfully reduced the hybridization length to 2 h.

Acknowledgments

Most of the work concerning embryo staging through whole-mount in situ hybridization was done with the support of INRA, EU and ANR grants and/or DFG grants, as specified in each of our publications, but also with the funding for French-German exchanges through DAAD and PHC (Procope). We also gratefully acknowledge the colleagues who helped us publishing nice WISH data (in Bonn: S. Molly and H. Langmann, in Jouy: E. Campion) and having access to embryos (in Jouy: J. Marchal, C. Richard).

References

1. Downs KM, Davies T (1993) Staging of gastrulating mouse embryos by morphological landmarks in the dissecting microscope. Development 118:1255–1266
2. Hamburger V, Hamilton HL (1992) A series of normal stages in the development of the chick embryo. 1951. Dev Dyn 195:231–272
3. Puschel B, Blum M, Viebahn C (2010) Whole-mount in situ hybridization of early rabbit embryos. Cold Spring Harb Protoc 2010(1):pdp prot5355
4. Weisheit G, Mertz D, Schilling K, Viebahn C (2002) An efficient in situ hybridization protocol for multiple tissue sections and probes on miniaturized slides. Dev Genes Evol 212:403–406
5. Hue I, Renard JP, Viebahn C (2001) Brachyury is expressed in gastrulating bovine embryos well ahead of implantation. Dev Genes Evol 211:157–159
6. Idkowiak J, Weisheit G, Plitzner J, Viebahn C (2004) Hypoblast controls mesoderm generation and axial patterning in the gastrulating rabbit embryo. Dev Genes Evol 214:591–605
7. Idkowiak J, Weisheit G, Viebahn C (2004) Polarity in the rabbit embryo. Semin Cell Dev Biol 15:607–617
8. Guillomot M, Turbe A, Hue I, Renard JP (2004) Staging of ovine embryos and expression of the T-box genes Brachyury and Eomesodermin around gastrulation. Reproduction 127:491–501
9. Hassoun R, Schwartz P, Feistel K, Blum M, Viebahn C (2009) Axial differentiation and early gastrulation stages of the pig embryo. Differentiation 78:301–311
10. Acloque H, Wilkinson DG, Nieto MA (2008) In situ hybridization analysis of chick embryos in whole-mount and tissue sections. Methods Cell Biol 87:169–185
11. Brend T, Holley SA (2009) Zebrafish whole mount high-resolution double fluorescent in situ hybridization. J Vis Exp 25:25
12. Degrelle SA, Le Cao KA, Heyman Y, Everts RE, Campion E, Richard C, Ducroix-Crepy C, Tian XC, Lewin HA, Renard JP, Robert-Granie C, Hue I (2011) A small set of extra-embryonic genes defines a new landmark for bovine embryo staging. Reproduction 141:79–89
13. Gawantka V, Pollet N, Delius H, Vingron M, Pfister R, Nitsch R, Blumenstock C, Niehrs C (1998) Gene expression screening in Xenopus identifies molecular pathways, predicts gene function and provides a global view of embryonic patterning. Mech Dev 77:95–141
14. Weiszmann R, Hammonds AS, Celniker SE (2009) Determination of gene expression patterns using high-throughput RNA in situ hybridization to whole-mount Drosophila embryos. Nat Protoc 4:605–618
15. Diez-Roux G, Banfi S, Sultan M, Geffers L, Anand S, Rozado D, Magen A, Canidio E, Pagani M, Peluso I, Lin-Marq N, Koch M, Bilio M, Cantiello I, Verde R, De Masi C, Bianchi SA, Cicchini J, Perroud E, Mehmeti S, Dagand E, Schrinner S, Nurnberger A, Schmidt K, Metz K, Zwingmann C, Brieske N, Springer C, Hernandez AM, Herzog S, Grabbe F, Sieverding C, Fischer B, Schrader K, Brockmeyer M, Dettmer S, Helbig C, Alunni V, Battaini MA, Mura C, Henrichsen CN, Garcia-Lopez R, Echevarria D, Puelles E, Garcia-Calero E, Kruse S, Uhr M, Kauck C, Feng G, Milyaev N, Ong CK, Kumar L, Lam M, Semple CA, Gyenesei A, Mundlos S, Radelof U, Lehrach H, Sarmientos P, Reymond A, Davidson DR, Dolle P, Antonarakis SE, Yaspo ML, Martinez S, Baldock RA, Eichele G, Ballabio A (2011) A high-resolution anatomical atlas of the transcriptome in the mouse embryo. PLoS Biol 9(1):e1000582
16. Degrelle SA, Campion E, Cabau C, Piumi F, Reinaud P, Richard C, Renard JP, Hue I (2005) Molecular evidence for a critical period in mural trophoblast development in bovine blastocysts. Dev Biol 288:448–460
17. Shimizu H, Uchibe K, Asahara H (2009) Large-scale whole mount in situ hybridization of mouse embryos. Methods Mol Biol 577: 167–179
18. Puschel B, Bitzer E, Blum M, Viebahn C (2010) Mounting, embedding, and sectioning of early rabbit embryos. Cold Spring Harb Protoc 2010(1):pdp prot 5356

19. Degrelle SA, Jaffrezic F, Campion E, Le Cao KA, Le Bourhis D, Richard C, Rodde N, Fleurot R, Everts RE, Lecardonnel J, Heyman Y, Vignon X, Yang X, Tian XC, Lewin HA, Renard JP, Hue I (2012) Uncoupled embryonic and extra-embryonic tissues compromise blastocyst development after somatic cell nuclear transfer. PLoS One 7:e38309

Chapter 11

Chicken Embryonic Stem Cells: Establishment and Characterization

Pauline Aubel and Bertrand Pain

Abstract

Embryonic stem (ES) cells are unique models for investigating early development and cell differentiation. First identified in mouse and later in other mammals, these cells have also been isolated in avian species. Here, using chicken as a model, we describe a set of protocols allowing the isolation, maintenance, genetic modification, differentiation, and injection of the chicken embryonic stem (cES) cells into embryos for obtaining chimeric animals.

Key words Chicken, Embryo, Embryonic stem cells, Blastoderm cells, Chimera, Differentiation, Embryoïd body, Transfection, Electroporation, Lentiviral infection

1 Introduction

Embryonic stem (ES) cells were isolated in the early 1980 from mouse (mESC) and in the late 1990 from primate and human. These cells present the unique property of self-renewal and the ability to generate differentiated progeny of all embryonic lineages both in vitro and in vivo. In mouse, ES cells contribute to both somatic and germinal lineages once injected into a recipient embryo at the blastocyst stage [1].

In avian species, chicken embryonic stem (cES) cells were isolated and grown for long-term culture from in vitro cultured early chicken blastodermal cells (cBC). cES cells can be maintained under specific culture conditions and have been characterized on the basis of their morphology, biochemical features, in vitro differentiation potential, and in vivo morphogenetic properties [2, 3]. They have been genetically modified using various vectors including expression vectors [4], gene trap vectors [5], homologous recombination vectors [6], conditional shRNA expressing vectors [7], as well as retroviral and lentiviral vectors (unpublished data). Using a differential expression screen, the avian homologues of the

Ramiro Alberio (ed.), *Epiblast Stem Cells: Methods and Protocols*, Methods in Molecular Biology, vol. 1074, DOI 10.1007/978-1-62703-628-3_11, © Springer Science+Business Media New York 2013

OCT4 (*cPOUV*) and *NANOG* mammalian genes were identified and cloned. Both genes are involved in the maintenance of chicken pluripotency [7].

Here, we present the main protocols for the isolation, maintenance, genetic modification, differentiation, and injection of the cES cells into embryos for generating chimeric animals.

2 Materials

2.1 Eggs

1. Eggs: fertile freshly laid eggs just collected. The eggs can be stored at 16 °C for 2–5 days before being processed. Eggs from meat strains of chicken provide more efficient derivation of cES cells comparatively to egg laying strains.

2.2 Culture Media

1. ESA complete medium (ESA CM): in a 500 mL DMEM/F12 bottle, add 50 mL of ES tested batch of Fetal Bovine Serum (FBS), non-heat inactivated, 6 mL NEAA (Non Essential Amino Acid) stock solution, 6 mL 100 mM Sodium Pyruvate stock solution, 6 mL 10,000 U/1,000 U Penicillin/Streptomycin stock solution, 6 mL 200 mM glutamine stock solution, 7 μL 14 M β-Mercaptoethanol stock solution. Store ESA CM medium at 4 °C for up to 2 weeks.
2. ESA Proliferative Medium (ESA PM): add 5 ng/mL IGF 1, 1 ng/mL SCF, 1 ng/mL IL6, 1 ng/mL sIL6 Rα, and 1,000 U/mL LIF in the ESA CM bottle. All the factors are recombinant either from human or mouse origin.
3. ESA Differentiation Medium (ESA DM): same as for ESA CM but with only 25 mL of FBS instead of 50 mL and no factors added.
4. STO medium: in a 500 mL DMEM/F12 bottle, add 25 mL FBS, 6 mL 10,000 U/1,000 U Penicillin/Streptomycin stock solution, 6 mL 200 mM glutamine stock solution. Store at 4 °C for up to 2 weeks.
5. Freezing Medium: 80 % FBS, 20 % DMSO volume/volume. Keep on ice and prepare extemporaneously.
6. 0.1 % gelatin solution: dissolve and autoclave 0.5 g bovine skin gelatin per 500 mL PBS 1×. Store at room temperature (RT). For the gelatin coating, add the 0.1 % gelatin solution at RT for at least 1 h using around 0.5 mL/cm^2 dish (i.e., 5 mL per 100 mm dish). Remove the solution before cell platting. Dishes, wells, slides can be coated in advance and kept dried and sterile for few days at RT once gelatin has been removed.

2.3 Disposable Sterile Plastic

1. 50 mL centrifuge tubes.
2. Cell culture-treated plates (6 wells, 12 wells, 24 wells), and 100 mm dish.

3. Insulin syringe (1 mL).
4. Large 2 L Becker Tank.
5. Micropipettes and sterile tips.
6. Pipettes (25, 10, 5 mL).
7. Plastic Bacteria Petri dishes (150 mm).
8. Eppendorp tube (1.5 mL).

2.4 Dissection Material

1. Borosillicate 20 μL capillary (Drummond).
2. KovaR slide.
3. Microcaps (Drummond).
4. Micropipettes and filtered tips.
5. Plastic loop on wooden sticks (*see* **Note 1**), autoclaved.
6. Punched Paper ring in 3M Whatman paper—sterilized 12 min at 180 °C.
7. Round Glass cover slips (12 mm diameter).
8. Small curved dissecting forceps.
9. Small dissecting forceps.
10. Small straight surgical scissors.

2.5 Chemical Reagents and Antibodies (See Note 2)

1. ATRA (All *trans* retinoic acid). Dissolve 50 mg in 15 mL of absolute ethanol to get a 5×10^{-2} M stock solution. Aliquot, protect from light and store at −20 °C. Dilutions performed in absolute ethanol.
2. DMSO (dimethylsulfoxide) anhydrous cell culture tested quality. Store at 4 °C in aliquots.
3. EMA1 and SSEA1 antibodies: perform small aliquots by diluting these IgM antibodies in glycerol vol/vol. Store at −20 °C.
4. Fixative solution: 1.5% Formaldehyde, 0.5% Glutaraldehyde in PBS. Keep at 4 °C.
5. Mitomycin C: dissolve 2 mg in 1 mL PBS leading to a 500× stock solution. Keep at −20 °C in aliquots and protect from light.
6. Paraformaldehyde solution: dissolve 20 g paraformaldehyde at 68 °C heated PBS. Allow to cool, aliquot in 50 mL tube, and store at −20 °C.
7. Quaternary ammonium compound (QAC).
8. Sodium hydroxide solution (10 N): dilute to 0.2 N for cleaning the plastic loop.

2.6 Cell Culture Laboratory Environment

1. Bench coat.
2. Bench flux.
3. Centrifuge.
4. CO_2 gas incubator.

5. Gloves.
6. Inverted Microscope.
7. Laboratory coat.

3 Methods

3.1 Blastoderm Collection

1. Disinfect the eggs with quaternary ammonium solution and then wiped with 70 % ethanol.
2. Break the shell neatly and separate the white and yolk carefully while the embryo is located at the surface of the yolk (Fig. 1a).
3. Pour an average of 5–8 yolks in a 150 mm dish, with the embryo directly placed upward.
4. Use for each one small punched ring of sterile 3M paper to remove the excess of white at the surface of the embryo.
5. Place then a new ring paper at the surface of the yolk with the dissecting forceps, the small hole surrounding the embryo (Fig. 1b).

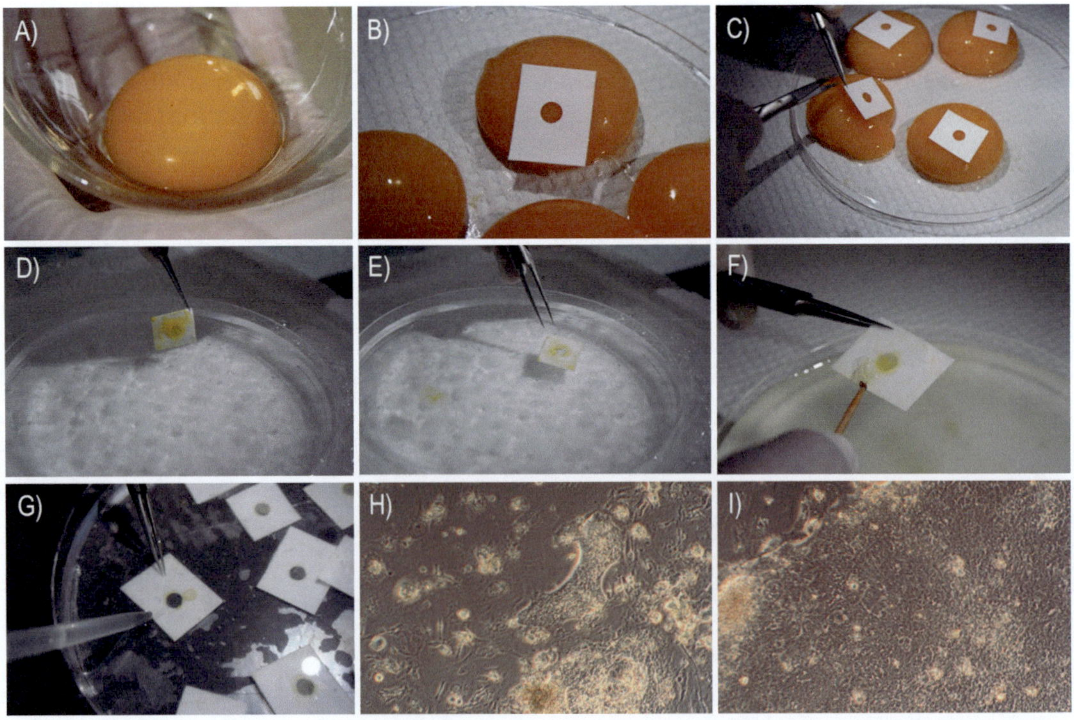

Fig. 1 Yolk egg obtained from freshly harvested fertile egg with the embryonic disc at the top (**a**). The paper ring is placed at the top of the yolk, the small hole surrounding the embryo (**b**). Cut rapidly the yolk with scissors to obtain the paper ring with the embryo (**c**). Submerge the paper ring in cold PBS (**d**), rinse it in fresh PBS (**e**), and clean it with the plastic loop (**f**). Detach the cells with a pipette (**g**). After plating, clusters of small round cells appear on the inactivated STO feeder cells (**h**) and can be passaged to obtain more homogenous cES cells (**i**)

6. After 1–3 min (depending on the dryness of the laboratory atmosphere) to allow the paper to stick at the surface, perform 3–4 small incisions around the ring (Fig. 1c).
7. Take rapidly the paper ring bearing the embryo in its center with the small curved dissecting forceps.
8. Immerge it carefully in cold PBS by taking care to enter the ring vertically to the liquid surface to avoid the embryo to detach itself (Fig. 1d).
9. Place the ring with the yolky side upward in the PBS (Fig. 1e).
10. Collect several embryos in parallel.

3.2 Blastodermal Cell Preparation and Initial Plating

1. Transfer the embryos on the paper ring to a new dish with fresh cold PBS.
2. Remove gently the excess of yolk using the loop by taking care to maintain the embryo on the paper ring (Fig. 1f). Once cleaned, the embryo is placed in another fresh dish containing PBS.
3. Detach the embryo with a 1 mL pipette tip from the paper ring (Fig. 1g) and place it in 0.25 mL of ESA PM complete medium in an Eppendorp tube.
4. Dissociate the cells mechanically by several up and down gentle pipetting.
5. Plate the cells directly in a well of a 12-well plate with inactivated STO feeder (cf infra Subheading 3.3) containing 0.75 mL of ESA PM complete medium.
6. Plate one embryo per well and incubate in the incubator at 39 °C under 10 % CO_2.
7. Add 0.5 mL fresh ESA PM medium the following day.
8. Rinse the cells twice with PBS 2 days after the initial plating once the cells attached.
9. Add fresh ESA PM medium.
10. Dissociate each well when the cells emerge—usually from day 3 to day 5.
11. Plate all the cells in 4 mL ESA PM medium in a 6-well plate on STO irradiated feeder cells (cf Subheading 3.3).
12. Wells in which colonies and foci are small, round cells will be selected for further dissociation and amplification (Fig. 1h). Not before the fourth to fifth passage will these cells be considered as established, and then grow more homogenously (Fig. 1i).
13. Freeze stocks of early passages cES.
14. Maintain the cells routinely (cf Subheading 3.4).

3.3 Feeder Production

1. Maintain the STO cells routinely at 37 °C under 7.5 % CO_2 in STO medium.
2. Change the medium every 2 days.
3. Pass the cells by trypsinization every 3–4 days.
4. For maintenance and amplification, plate STO cells at 1×10^6 cells per 100 mm dish in 12 mL STO medium.
5. For Mitomycin C inactivation, plate STO cells at 1×10^6 cells per 100 mm dish in 12 mL STO medium.
6. The next day, treat the STO for 3 h with 4 μg/mL Mitomycin C in 5 mL STO medium (*see* **Note 4**).
7. After treatment, use the Mitomycin C STO-treated cells for up to 4–5 days.
8. For irradiation, wash the dissociated STO cells in PBS, count and resuspend at 5×10^6 cells per mL in STO medium.
9. Irradiate the STO cells at 50 Gy and either plate them immediately, or freeze them.
10. For liquid nitrogen storage, freeze the irradiated STO cells at 5×10^6 cells per mL per vial in 50 % STO medium 50 % freezing medium vol/vol.
11. For direct plating, or thawed irradiated frozen cells, plate at 1×10^6 cells per 100 mm dish or 2.5×10^5 cells per well of a 6-well plate in STO medium.
12. For irradiation control, plate an aliquot of 1×10^6 cells of each irradiated batch in one 100 mm culture dish and keep under standard STO culture condition for 1 month with a medium change every 5 days. After 1 month, rinse the dish with PBS, fix for 10 min by 5 mL methanol, and stain with Wright Giemsa solution for 10 min before rinsing the dish. No colonies or foci should be present for a batch to be validated.

3.4 cES In Vitro Maintenance

Once established, 4–5 passages after the initial plating, the cES cells are routinely maintained on STO irradiated feeder cells in a CO_2 incubator at 39 °C under 10 % CO_2.

1. Change the ESA PM medium every day.
2. Passage the cells every 3 days.
3. Collect 5 mL of conditioned medium in a 50 mL tube.
4. Rinse the cells twice with PBS and discard.
5. Add Accutase (0.5 mL/well, 2 mL/100 mm dish).
6. Incubate the cells at RT for 2–3 min until the cells detach.
7. Check the dissociation under the microscope.
8. Collect the cells with a 10 mL pipette in the 50 mL tube containing the 5 mL of conditioned medium.
9. Rinse twice the dish with 5 mL PBS and collect in the same tube.

10. Homogenize the cell suspension by gentle up and down pipetting.
11. Take a 50 μL aliquot.
12. Mix vol/vol with Trypan blue for counting (*see* **Note 2**).
13. Count the cells on a Kova slide.
14. Centrifuge 5 min at 300 × *g* at RT.
15. Resuspend the cell pellet in fresh medium at the right concentration according to the cell use.
16. For routine maintenance, plate 5×10^5 cells per 100 mm dish, 2×10^5 in a 6-well plate on irradiated STO feeder cells (ratio 1:2).

3.5 Freezing the cES

1. Dissociate and count the cells as described.
2. Centrifuge and resuspend the cell pellet in 1 mL per 10×10^6 cells.
3. Check the final volume of cell suspension.
4. Add slowly, drop by drop, the same volume of ice-cold Freezing medium stirring gently between each drop but avoiding any bubbles.
5. Place the tube on ice.
6. Distribute 1 mL of cell suspension (5×10^6 cells) per frozen vial placed on ice.
7. Close the vials tightly.
8. Allow the vials to be cooled at −80 °C in a cooling box overnight.
9. Transfer the frozen vials in liquid nitrogen for long-term storage (*see* **Note 3**).

3.6 Thawing

1. Prepare a water bath at 37 °C.
2. Remove the frozen vial from liquid nitrogen (*see* **Note 3**).
3. Immerge the bottom of the frozen vial in the water bath.
4. Wait until the cell suspension melts but not completely.
5. Wipe the vial with 70 % ethanol.
6. Open the vial and transfer it on ice.
7. Remove the cell suspension with a 1 mL micropipette tip and place it in a 50 mL tube.
8. Rinse the vial with 1 mL of cold ESA PM medium.
9. Add drop by drop this 1 mL on the thawed cell suspension stirring gently between each drop.
10. Add 1 mL of ESA PM cold medium in the same condition.
11. Continue to add 6 mL twice.
12. Centrifuge 10 min at 150 × *g* at RT.
13. Resuspend the cell pellet in 8 mL of ESA PM medium.

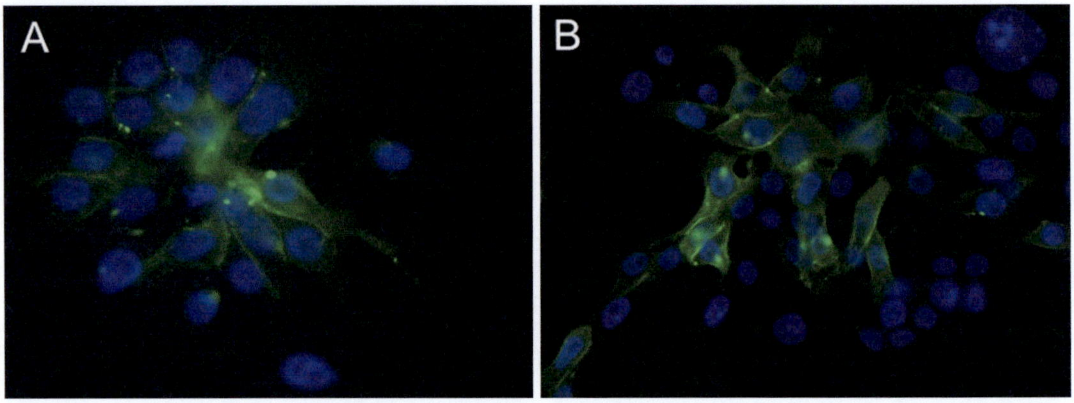

Fig. 2 cES cells are positive for SSEA1 (**a**) and EMA1 (**b**) antibodies. Cells were grown on coverslips in ESA PM medium, and labelled for those membrane bound antigens that are specific to ES cells

14. Plate in 2 wells of a 6-well plate onto STO irradiated feeder cells.
15. Incubate at 39 °C under 10 % CO_2 and wait 24–48 h before changing the medium.
16. Passage the cells as described and maintain them routinely.

3.7 Antibody Characterization

1. Prepare sterile glass coverslips.
2. Add one coverslip in a well of a 24-well plate.
3. Gelatin coat the coverslip.
4. Plate irradiated STO cells on gelatin-coated coverslip.
5. After dissociation, plate 5×10^4 cES on STO-coated glass coverslip in ESA–PM medium.
6. Allow the cells to grow 1–2 days to reach sub-confluence.
7. After rinsing the cells twice with PBS, fix the cells 15 min at 4 °C with Fixative solution.
8. Proceed with the antibody staining.

Typical SSEA1 and EMA1 stainings are illustrated in Fig. 2a, b.

3.8 Embryoïd Body (EB) Formation

Embryoid body formation allows the cell to differentiate in the different embryonic lineages. Depending on the experimental design, the subsequent culture of the EBs will be either stopped at a defined day or continued for extra few days to get a kinetic of differentiation.

1. After dissociation, resuspend the cES pellet in ESA DM medium.
2. Adjust the cell concentration to 7×10^4 cells per mL.
3. Align 8 series of 30 μL drops of cell suspension on the cover of the large bacterial dish using the multichannel pipette (around 2,000 cells per hanging drop).

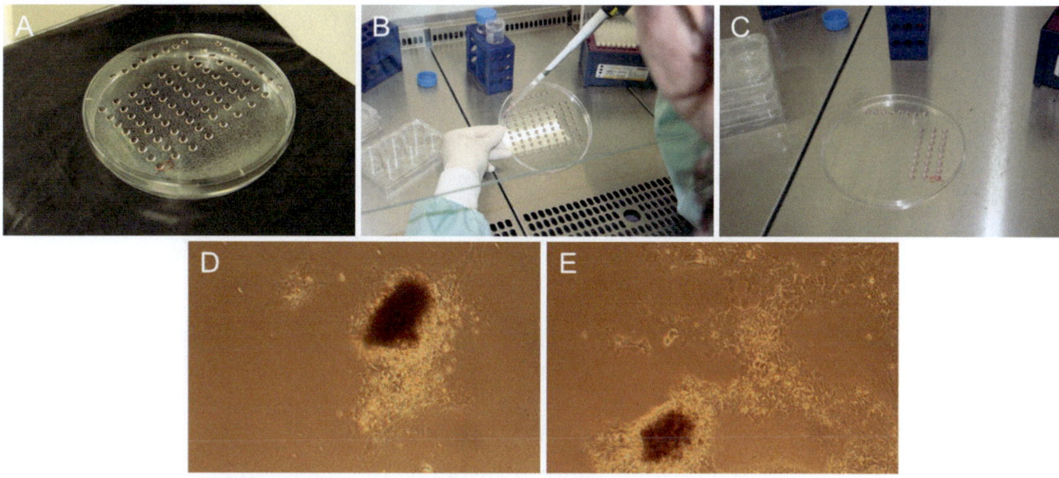

Fig. 3 Embryoid bodies (EBs) are formed by allowing the cES to grow in hanging drops in 30 μL of ESA DM medium (**a**). After 3 days, the EBs are harvested (**b**, **c**) by adding ESA DM medium to the drops and collecting the EBs. EBs are platted and induced further to differentiate (**d**, **e**) in the presence or absence of various inducers

4. Pour sterile distilled water in the bacterial dish.
5. Return the cover (Figure 3A).
6. Incubate at 39 °C, 10 % CO^2 for 3 days.
7. After 3 days, return the cover again.
8. Incline the lid and add 200 μL of ESA DM medium at the top of each line (Fig. 3b).
9. Collect the EBs from one line by a 1 mL pipette tip (Fig. 3c).
10. Plate directly the EB in ESA DM medium in wells treated with gelatin, laminin, or fibronectin depending on the experimental plan.
11. Allow the EB to attach for 2–3 days (Fig. 3d, e).
12. Change medium carefully and add the inducer such as Retinoic acid at 5×10^7 M (*see* **Note 2**), DMSO at 1 % vol/vol or other factors or chemicals.
13. Change the medium every 2 days for 5–7 days according to the experimental plan.
14. Perform the analysis of the differentiated cells spreading from the attached EBs by real-time PCR and/or direct immunofluorescence.

3.9 In Vitro cES Genetic Modification by Transfection (See Note 4)

1. On day 0, plate 1×10^5 cES cells per well of a 6-well plate in 4 mL of ESA PM medium.
2. On day 1, remove the medium, rinse the cells with PBS, and add 2 mL of serum-free medium (*see* **Note 5**).
3. Add 2 μg of plasmid to the transfection mixture (*see* **Note 6**).

4. Add the transfection mixture for 6 h and incubate the cells.
5. Remove the medium and add 4 mL of fresh of ESA PM medium.
6. On day 2, change the medium.
7. On day 3 and 4, change the medium.
8. On day 5, analyze the cells for transient expression or add the selection for getting stable clones if the vector allows it (*see* **Note 7**).
9. After selection, pick the clones individually or pool them for further amplification and analysis.

3.10 In Vitro cES Genetic Modification by Electroporation

1. Dissociate and count the cells.
2. Centrifuge and resuspend the cell pellet in a serum-free medium at $1–2 \times 10^6$ cells per mL (*see* **Note 5**).
3. Centrifuge again and resuspend at 6.25×10^6 cells per mL in serum-free medium.
4. Mix 0.8 mL of cells (corresponding to 5×10^6 cells) with 15 μg of linear plasmid.
5. Place the mix in an electroporation cuvette.
6. Wait for 10 min at RT.
7. Electroporate at 270 V, 700 μF (BioRad electroporator).
8. Place the cuvette on ice for 10 min.
9. Plate the cells in two 100 mm dishes on STO irradiated feeder cells in 10 mL ESA PM medium.
10. After 48 h of the electric pulse perform the analysis, or add the selection if the vector allows it.

3.11 In Vitro cES Genetic Modification by Lentiviral Infection (See Note 8)

1. On day 0, plate 1×10^5 cells per well of a 6-well plate.
2. On day 1, rinse the cells with PBS and add 2 mL of fresh ESA PM medium.
3. Add Polybrene at 6 μg/mL with the viral suspension using a multiplicity of infection (m.o.i.) from 10 to 50 according to the expected process.
4. Incubate at 37 °C, 7.5 % CO_2 for 4 h.
5. Remove the medium.
6. Add 5 mL of fresh ESA PM medium and incubate the cells at 39 °C, 10 % CO_2.
7. On day 2, rinse the cells with PBS and add 5 mL fresh ESA PM medium.
8. On day 4, analyze directly the cells or add the selection if the vector allows it.

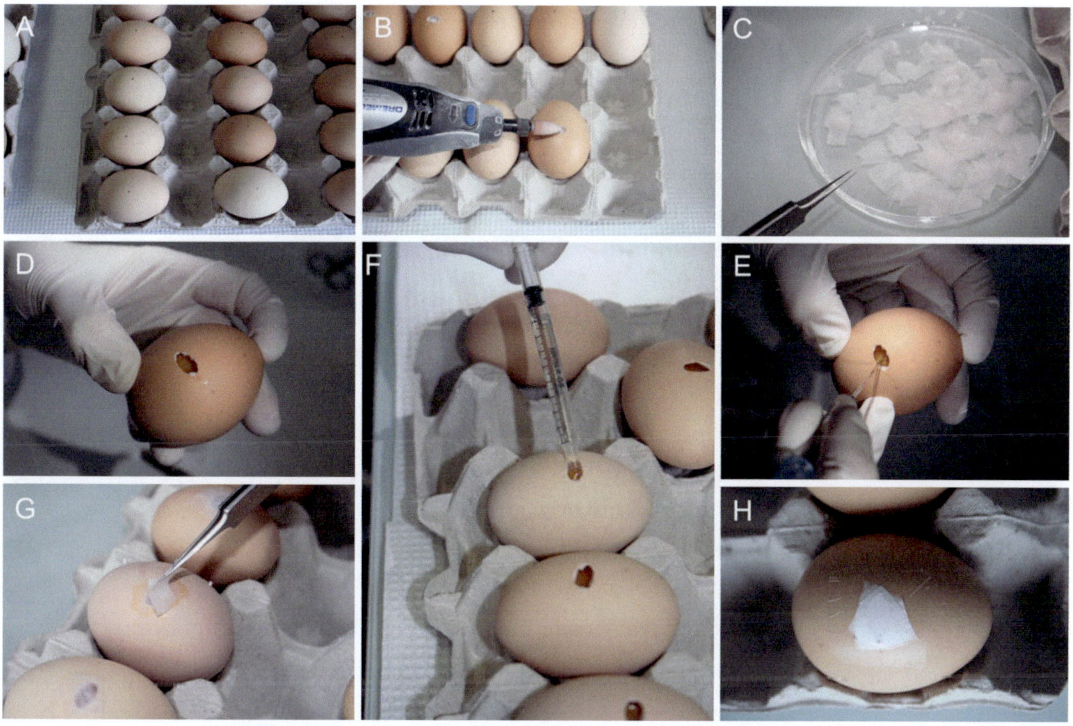

Fig. 4 Eggs are put horizontally and labeled (**a**) for allowing the embryo to place itself just beneath the top of the egg. The egg shell is then drilled (**b**), the membrane shell cut (**d**) for a direct injection of the dissociated cells into the recipient embryo (**e**). After injection, the eggs are filled with white egg albumen (**f**). With previously prepared membrane shells (**c**), the window is closed (**g**) before being sealed with adhesive dressing for incubation (**h**)

3.12 In Vivo Injection

One of the interest of the cES is their ability to contribute to embryo when injected into recipient embryo. As an example, using GFP labeled cells, the chimerism can be followed by microscopic observation in early embryos.

1. Place freshly laid fertile eggs horizontally on an egg tray at 14–16 °C the previous day of the injection, up to 2–3 days before the injection.
2. Label the horizontal position on the upside of the egg (Fig. 4a).
3. Remove the white and thick shell membranes from spared egg shells.
4. Cut these shell membranes in small squares of around 1 cm^2 (Fig. 4c).
5. Keep them in cold PBS containing 10 % liquid white egg albumen and 1 % Penicillin–Streptomycin (*see* **Note 9**).
6. Prepare a bacterial dish containing white egg albumen—for dropping the membranes before their use.

7. On the day of injection, dissociate and count the GFP-labeled cES cells.
8. Centrifuge and resuspend the cell pellet at 1×10^6 cells per mL in ESA PM medium.
9. Keep the cells on ice.
10. On the day of injection, disinfect the eggs with quaternary ammonium solution and wipe them individually with 70 % ethanol.
11. Drill a small window in the shell on the top of the labeled mark by avoiding damaging the shell membrane just beneath the shell (Fig. 4b).
12. Wipe again the drilled eggs with 70 % ethanol (to remove the shell dust).
13. Keep them horizontally.
14. Place one 100 μL drop of cell suspension in bacterial dish under a binocular.
15. Adapt the sterile injection capillary in the injection set.
16. Load 10–12 μL of cell suspension in the capillary.
17. Cut the membrane shell of the drilled egg with a scalpel blade and remove with the small dissecting forceps the membrane to get a clear window (Fig. 4d).
18. Inject 2–3 μL into the subgerminal cavity of the embryo just beneath the drilled window (Fig. 4e) (*see* **Note 10**).
19. Place the injected egg horizontally again.
20. Fill the injected eggs with liquid white egg albumen using a 1 mL insulin syringe (Fig. 4f).
21. Close the small window by two shell membrane pieces previously dropped in liquid white albumen (Fig. 4g).
22. Allow the shell membranes to dry for 1–2 h depending on the air dryness (*see* **Note 11**).
23. Add a sealing dressing to avoid leaks and dehydration during incubation (Fig. 4h).
24. Incubate normally in an egg incubator.

3.13 Embryo Analysis

Chimeric embryos can be obtained with GFP positive cES cells (Fig. 5a, b).

1. Crack the injected eggs in a bacteria dish after the expected incubation following the experimental plan.
2. Remove the embryo by producing a small incision using scissors.
3. Take the embryo by the enveloping membranes.
4. Rinse it several times in PBS.

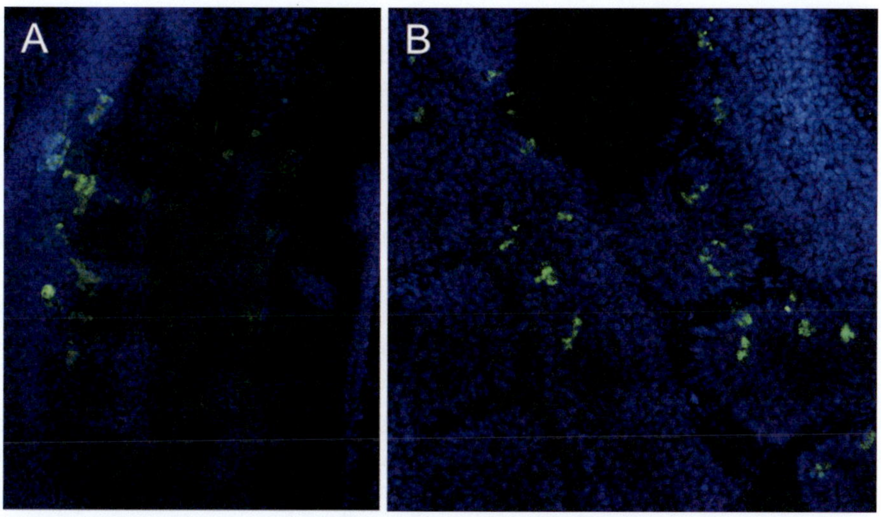

Fig. 5 After few days of development, the GFP-labeled cells can be detected in small clusters (**a**) or more individual cells (**b**) in different parts of the embryo. Nuclei have been labeled with Hoescht 33342

5. Fix the embryos either in Fixative solution or in Paraformaldehyde solution depending on the experimental design.
6. Analyze according to the experimental plan.

4 Notes

1. The loops are just made using a small hood stick at the top of which a loop of thin plastic wire is sealed. After each preparation the loops can be cleaned using a NaOH 0.2 N solution, before being rinsed extensively in water and autoclaved.
2. The chemicals present different level of toxicity and some are harmful products. To be handled with great care and according to the regulation.
3. Liquid nitrogen requires specific regulation and care for its storage and proper manipulation.
4. cES can be genetically modified either by transfection, electroporation, or viral infection.
5. Serum-free medium is routinely either Opti MEM (Invitrogen) or simply DMEM/F12 medium, but others could also work.
6. The transfection mixture usually contains 0.2 mL serum-free medium, 2 μg plasmid, and the liposome. Toxicity and efficiency could be different between different brand of liposomes. Among those presenting the best ratio, it could be mentioned the Fugene (Roche) and the Lipofectamin (Invitrogen) compounds. Other could also work efficiently.

7. For selection of the cES, Neomycin (G418) is added at 250 μg/mL for 7–8 days, Hygromycin at 75 μg/mL, for 5–7 days, or Puromycin at 1 μg/mL for 5–7 days.
8. Lentiviral manipulation requires specific regulation and care.
9. The preparation of shell membranes can be performed in advance as those membranes, once cut, are easily kept at −20 °C in a 50 mL tube in the PBS 10 % liquid white egg albumen 1 % Penicillin–Streptomycin stock solution.
10. Some embryos will not be found just beneath the opened window. If clearly visible, they can be injected, otherwise discard the egg as it is often associated with a bad quality embryo.
11. The shell membranes dry when in contact with the air and become white. According to the dryness of the air, the drying time varies. Pay attention to the dehydration of the injected embryo in a very dry atmosphere.

Acknowledgments

This work was supported by grants from ANR (ANR 09 BLAN 0141 01).

We thank the laboratory members for their help and comments.

References

1. Chambers I, Smith A (2004) Self renewal of teratocarcinoma and embryonic stem cells. Oncogene 23:7150–7160
2. Pain B, Clark ME, Shen M et al (1996) Long term in vitro culture and characterisation of avian embryonic stem cells with multiple morphogenetic potentialities. Development 122:2339–2348
3. Petitte JN, Liu G, Yang Z (2004) Avian pluripotent stem cells. Mech Dev 121:1159–1168
4. Pain B, Chenevier P, Samarut J (1999) Chicken embryonic stem cells and transgenic strategies. Cells Tissues Organs 165:212–219
5. Acloque H, Risson V, Birot AM et al (2001) Identification of a new gene family specifically expressed in chicken embryonic stem cells and early embryo. Mech Dev 103: 79–91
6. Pain B, Samarut J, Valarche I et al (2001) Exogenous protein expression system in avian system. Patent WO03043415
7. Lavial F, Acloque H, Bertocchini F et al (2007) The Oct4 homologue PouV and Nanog regulate pluripotency in chicken embryonic stem cells. Development 134: 3549–3563

Chapter 12

Manipulating the Avian Epiblast and Epiblast-Derived Stem Cells

Cantas Alev, Mikiharu Nakano, Yuping Wu, Hiroyuki Horiuchi, and Guojun Sheng

Abstract

Compared to eutherian mammals, birds retain a primitive form of epiblast development. Molecular studies of the avian epiblast can provide valuable insight for mammalian epiblast research. Here, we introduce several basic techniques in handling epiblast-stage embryos of the chick, the major model organism for avian developmental biology studies. We describe how to collect embryos for RNA extraction and gene expression analysis, to set up ex ovo New culture for overexpression, bead graft and small molecule-based inhibitor studies, and to carry out whole-mount RNA in situ hybridization analysis. We introduce a novel and simple method for molecular perturbation of the epiblast differentiation in ovo. We also describe how to perform primary chicken epiblast cell culture, to establish stable epiblast stem cell (Epi-SC) lines, and to assay for pluripotency in primary epiblast cells and Epi-SCs.

Key words Avian, Chicken, Epiblast, Stem cells, Pluripotency, New culture, Bead graft, In situ hybridization, Chicken ES cells, Epiblast stem cells

1 Introduction

Epiblast is a term used to describe the primitive ectoderm in amniotes (mammals, birds, and reptiles). In birds, it includes cells in both the area pellucida (equivalent to the mammalian epiblast) and the area opaca (similar to mammalian mural trophectoderm). Mammalian polar trophectoderm does not have an avian counterpart. Here, we will use the chick as the avian model and only concern ourselves with the area pellucida epiblast. We will refer to this part as the epiblast. It should be kept in mind that in addition to the three germ layers in the embryo proper, the epiblast in both the mammals and birds contributes to the extraembryonic mesoderm in its entirety and to the extraembryonic ectoderm partially.

The first day of chicken development takes place intrauterally. A fertilized and freshly laid hen's egg contains approximately

Ramiro Alberio (ed.), *Epiblast Stem Cells: Methods and Protocols*, Methods in Molecular Biology, vol. 1074, DOI 10.1007/978-1-62703-628-3_12, © Springer Science+Business Media New York 2013

40,000 cells and has reached the Eyal-Giladi and Kochav (EGK) stage X [1]. Intrauteral development (EGK I-X) and post-laying pre-streak development (EGK XI-XIV) are traditionally staged according to the EGK criteria [1] and correspond collectively to Stage 1 of the Hamburger and Hamilton (HH) system [2]. After the appearance of the primitive streak at HH2 (about 6–8 h of incubation post-laying), chicken development is staged according to the HH criteria [2]. Although the later half of the intrauteral development is relevant to issues concerning epiblast formation and pluripotency, technical limitations in obtaining these embryos preclude their routine use in labs without access to a poultry farm. In this chapter, we mainly discuss about how to handle chicken embryos during the first 18 h of its post-laying development (from stage EGK-X up to stage HH5). Most pluripotency genes are markedly down-regulated in the epiblast by HH5 [3, 4]. This period of chicken development corresponds roughly to E5.5–E7.5 of mouse development.

2 Materials

2.1 Manipulation and Analysis of Epiblast-Stage Chicken Embryos

1. Pannett-Compton Solution: H_2O, 40 ml of solution-A, 60 ml of solution-B in 1 l final volume. Pannett-Compton Solution-A: H_2O, 121 g NaCl, 15.5 g KCl, 10.42 g $CaCl_2$–$2H_2O$, and 12.7 g $MgCl_2$–$6H_2O$ in 1 l final volume. Autoclave. Pannett-Compton Solution-B: H_2O, 1.88 g Na_2HPO_4, and 0.188 g NaH_2PO_4–$2H_2O$ in 1 l final volume. Autoclave.
2. 20× PBS: H_2O, 175.3 g NaCl, 22.7 g Na_2HPO_4, and 4.7 g NaH_2PO_4–H_2O in 1 l final volume. Autoclave. For 1× PBT: 1× PBS with 0.1 % Tween-20.
3. 4 % PFA: Add 4 % w/v of paraformaldehyde powder with continuous stirring to preheated (65 °C) 1× PBS. Adjust pH to 7.5 with 1 N NaOH. Cool to RT and add EGTA to final concentration of 2 mM. Alternatively, premade 4 % Paraformaldehyde Phosphate Buffer Solution can be purchased from commercial sources.
4. Postfix: 4 % formaldehyde and 0.1 % glutaraldehyde in 1× PBT.
5. Pre-hybridization solution: H_2O, 50 % formamide, 1.3× SSC, 5 mM EDTA, 50 μg/ml yeast RNA, 0.002 % Tween-20, 0.005 % CHAPS, and 100 μg/ml Heparin. All glassware and tools in making this solution should be baked. Aliquot into 50 ml tubes. Store at −20 °C.
6. 20× SSC: H_2O, 175.3 g NaCl, 88.2 g Na_3Citrate–$2H_2O$ in 1 l final volume. Adjust pH to 5.3 with citric acid.
7. 10× TBST: H_2O, 8 g NaCl, 0.2 g KCl, 25 ml of 1 M Tris–HCl pH 7.5, and 11 g Tween-20 in 100 ml final volume.

8. Color reaction solution: NTMT (1 ml) + NBT (2.3 μl) + BCIP (2.3 μl). Scale up accordingly. NBT: 4-nitro-blue tetrazolium chloride solution (Roche). BCIP: 5-bromo-4-chloro-3-indolyl-phosphate-toluidine salt solution (Roche). NTMT: H_2O, 1 ml of 5 M NaCl, 2.5 ml of 2 M Tris–HCl (pH 9.5), 1.25 ml of 2 M $MgCl_2$, 5 ml of 10 % Tween-20 in 50 ml final volume. Make fresh each time.
9. Affigel beads: Affi-Gel Blue beads (BioRad). AG1 ion exchange beads: AG1-X2 ion exchange beads (BioRad). Heparin-coated beads: Heparin immobilized on acrylic beads (Sigma).
10. Antibody blocking solution: 5 % heat-inactivated goat serum (30 min at 55 °C) and 1 mg/ml BSA in TBST.
11. pGEM-T Vector: pGEM-T Easy Vector System I (Promega).
12. Glass ring: Outer diameter 26 mm and inner diameter 23 mm. Custom made by local company (URIN, Kyoto, Japan).

2.2 Manipulation and Analysis of Epiblast Cells in Culture

1. Epiblast cell culture medium (ECC medium): KnockOut-DMEM (Invitrogen) supplemented with 20 % knockout serum replacement (KSR, Invitrogen), 2 % chicken serum, 1 mM sodium pyruvate, 1× nonessential amino acids, 2 mM L-glutamine, 1× nucleoside solution, 0.1 mM β-mercaptoethanol, 20 ng/ml recombinant chicken leukemia inhibitory factor.
2. 0.1 % gelatin solution: H_2O, 1 g gelatin in 1,000 ml final volume, followed by sterile filtration.
3. Ham-F12: F12 Nutrient Mixture.
4. Lipofectamine 2000 reagent: Lipofectamine™ 2000 Transfection Reagent (Invitrogen).
5. Mitomycin C solution: 1× PBS with 2 mg/ml mitomycin C.
6. 100× nucleoside solution: H_2O, 80 mg adenosine, 73 mg cytidine, 85 mg guanosine, 24 mg thymidine, and 73 mg uridine in 100 ml final volume, sterilized by passing through a 0.22 μm disposable filter.
7. Stem cell passaging tool: StemPro EZPassage (Invitrogen).
8. SuperScript™ III: SuperScript™ III First-strand Synthesis System (Invitrogen).
9. TOP10F *E. coli* strain: One Shot® TOP10 Chemically Competent *E. coli* (Invitrogen).
10. Zeocin: Zeocin™ Selection Reagent (Invitrogen).
11. ProBond Resin: ProBond™ Purification System with Antibody (Invitrogen).
12. pSecTag2A Vector: pSecTag2 Expression and Purification kit (Invitrogen).
13. Anti-chicken Nanog polyclonal rabbit antibody: supplied by Hiroshima University [5].

3 Methods

3.1 Obtaining Epiblast-Stage Chicken Embryos

Basic embryological precautions should be taken in staging and handing early-stage chicken embryos (*see* **Note 1**).

1. Prepare spoon, scissors, and coarse and fine forceps (Fig. 1a). Regular wash (soap, tap water, distilled water, and 70 % ethanol) of these tools is sufficient for most experiments.
2. Eggs are flipped upside down from their storage/incubation position. This way the yolk is positioned lower inside the egg. An incision is made with the tip of coarse forceps (Fig. 1b, c).
3. Top part of the egg shell is opened and removed (Fig. 1c, d) and most of the albumen removed (or collected in a small beaker if needed) (Fig. 1d, e).
4. Position the embryo at the center by gently moving the yolk with a spoon or the back of forceps. Do not touch the embryo directly. Use scissors to make 3–4 quick cuts around the embryo (Fig. 1e, f).
5. Scoop up the embryo, together with the vitelline membrane above and some yolk below, with a spoon. Dip the embryo into PBS or Pannett-Compton solution (Fig. 1g).
6. Remove the vitelline membrane (with embryo) from the yolk with fine forceps (Fig. 1h). Move slowly over the embryo area.
7. In most cases after more than a few hours of incubation, the embryo stays attached to the vitelline membrane (Fig. 1h). In cases when the embryo stays attached to the yolk, use the blunt end of a Pasteur pipette to suck up and transfer the embryo (Fig. 1i).
8. Use coarse forceps to hold down the vitelline membrane and fine forceps (with tips closed) to gently dislodge the embryo from the vitelline membrane by scraping the surface of the vitelline membrane towards to the embryo (Fig. 1j). This way the edges of the embryo are dislodged from the vitelline membrane, while most regions of the embryo are not touched by the forceps (Fig. 1k).
9. Once freed from the vitelline membrane (Fig. 1l), the embryo can be transported using the blunt end of a Pasteur pipette from dish to dish with minimal damage, so long as it does not come into the liquid/air interface. Excessive yolk can be removed with forceps or gentle blows using a Pasteur or Gilson pipette.
10. If needed (e.g., for RT-PCR or large-scale transcriptomic analysis), the epiblast region can be cut out from the embryo (Fig. 1m). In most cases, embryos will be fixed for later use. Fix them flat. This is very important. The epiblast is slightly

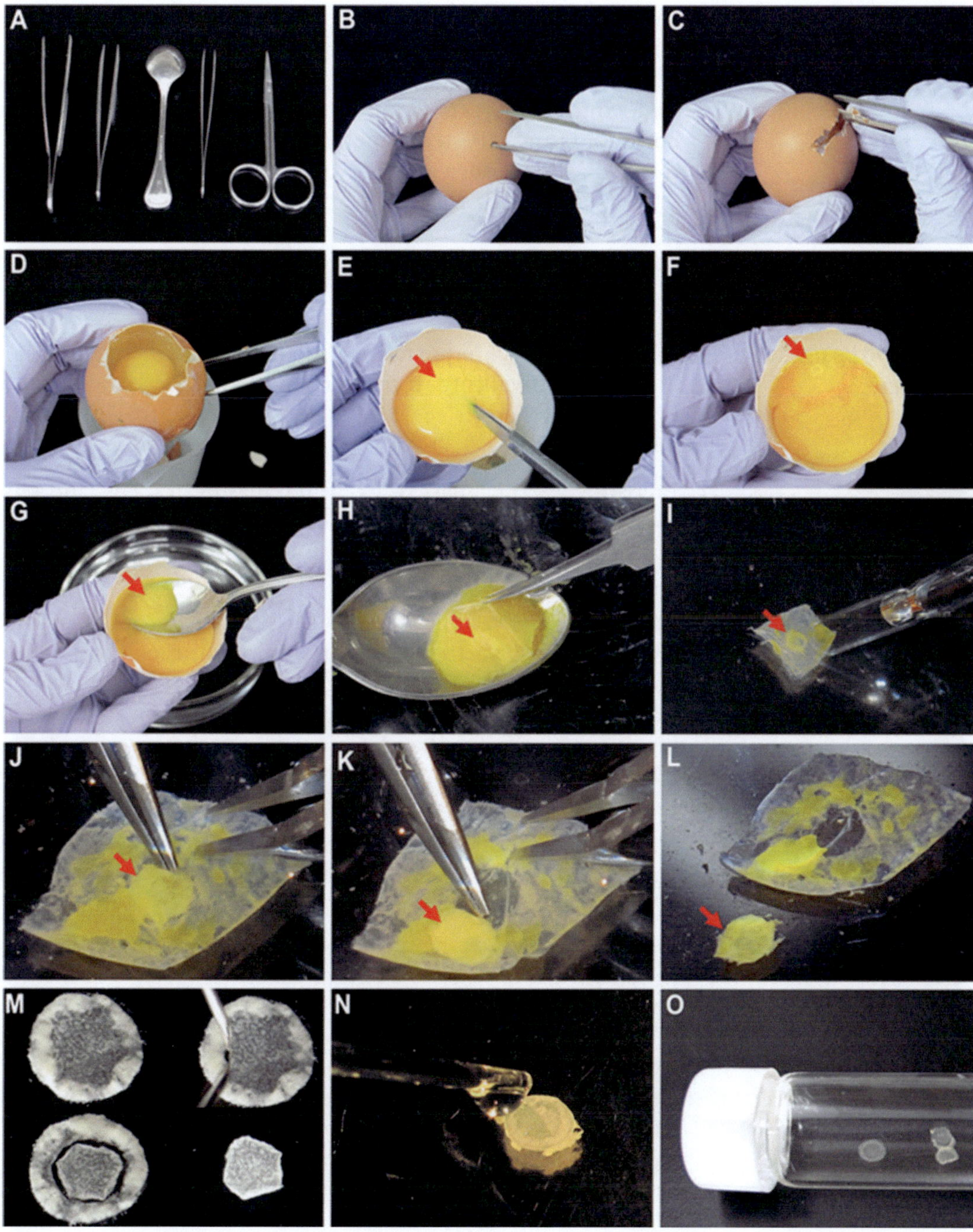

Fig. 1 Obtaining epiblast stage chick embryos. Unincubated eggs are opened (**a–d**) and embryos (*red arrow*) cut out and separated from the yolk as shown (**e-h**). Embryos are carefully detached from the vitelline membrane (**i–l**). The epiblast region can be cut out from freshly isolated embryos (**m**) or embryos can be fixed and used for subsequent analysis (**n**, **o**)

stretched in ovo, so it shrinks a bit in solution and can also fold up very easily. Flatten each embryo (it does not matter which side is facing up) with fine forceps, and further stretch it flat by removing all Pannett-Compton (or PBS) liquid around the embryo. Using a Pasteur pipette, gently add 1–2 drops of fixative (4 % PFA) on top (Fig. 1n). After 1–2 min, add a few drops. After 5–10 min of fixation, embryos can be handled quite roughly. With the blunt end of a Pasteur pipette, transfer embryos to a glass vial, fix for a couple of hours at RT or overnight at 4 °C (Fig. 1o).

3.2 Setting Up the New Culture

The New culture technique takes some practice to master. But it is the best ex ovo culture technique available for epiblast-stage chicken embryos and its versatility (*see* **Note 2**) justifies the initial steep learning curve.

1. Prepare glass tray, glass rings, watch glasses, forceps, and scissors (Fig. 2a). Regular wash of these tools (as in Subheading 3.1) is sufficient, but avoid using too much detergent and rinse with tap water thoroughly.
2. Pour 1–1.5 l of Pannett-Compton solution into a glass tray. Open eggs, collect thin albumen, remove as much thick albumen as possible without damaging the yolk. Pour intact yolk gently into the solution. Prepare a few yolks this way (up to 20 if you are experienced) and put several watch glasses inside the tray (Fig. 2b).
3. Use the back end of forceps to rotate the yolk so that the embryo is positioned on the top. Starting from one side of the yolk equator (Fig. 2c, e), cut all around the yolk with scissors (Fig. 2d, f).
4. Peel off the vitelline membrane with forceps gently and steadily (Fig. 2g). If the egg has been incubated for several hours, the embryo should stay attached to the vitelline membrane (Fig. 2f). If unincubated, the embryo tends to be detached from the vitelline membrane. In such cases, collect the embryo from the yolk using the blunt end of a Pasteur pipette, wash excessive yolk off the embryo and put the embryo back on to the vitelline membrane after the latter has been assembled. In the meantime, continue with the assembly of the vitelline membrane.
5. Move the vitelline membrane over a watch glass (yolky side up) (Fig. 2h) and position a glass ring on the membrane (Fig. 2i).
6. Use forceps to grab the membrane gently around the ring so that there is membrane margin all around the ring (Fig. 2j). Move the watch glass (with the embryo and the ring) out of the tray and on to the bench.

Fig. 2 Setting up New Culture. Egg yolks are collected in a glass tray filled with Pannett-Compton solution (**a–c**). Embryos attached to the vitelline membrane are obtained by generous circumferential cutting and careful dislodgement of the membrane plus embryos from the underlying yolk (**d–h**). The vitelline membrane (yolk side up) is carefully wrapped around and attached to a glass ring as shown (**i–m**). The assembled embryo is placed onto a 35 mm dish containing albumen (**o–r**). Albumen can be mixed with, e.g., small molecule-based inhibitors following successive dilution steps (**n**). Representative images of control and TGF-beta inhibitor (25 μM SB431542)-treated embryos reveal clear reduction in the expression of the pluripotency marker Nanog in the central epiblast upon New culture treatment in the presence of TGF-beta inhibition (**s**, **t**)

7. Reposition the ring as it may have shifted its position during the transfer. Use a Pasteur pipette to remove a bit of solution so that the top of the ring is exposed to the air. Use coarse forceps first, and then Pasteur pipettes (flame-polished to remove sharp edges) to grab the outer edges of the vitelline membrane over the ring (Fig. 2k). Because the top of the ring is exposed to the air, the membrane will (after 10–30 min) be attached to the ring after drying up. During this time window,

the membrane can be adjusted. An ideal outcome is that the vitelline membrane is taut under the ring and wraps around the top surface of the ring (Fig. 2l).

8. Cut off excessive vitelline membrane inside the ring (most lateral bits of the membrane) with scissors. Large pieces of yolk on the vitelline membrane inside the ring are removed with a Pasteur pipette (Fig. 2l, m). Always leave some liquid to cover the embryo. The assembly process may be stopped here for up to a couple of hours before moving on to the next step. Also at this step, the embryo can be removed for electroporation, detached embryos can be put back, and for bead graft, a few beads can be put onto the embryo.
9. Aliquot 2–3 ml of thin albumen in each 35 mm dish. For whole embryo treatment with small molecules, the chemical stock solution (or control solution), often dissolved in DMSO, is diluted in PBS first, followed by further dilution in the thin albumen (Fig. 2n). The volume of albumen in each dish can be reduced to 0.5–1 ml if cost of the chemical is a concern (e.g., SU5402).
10. Remove all Pannett-Compton solution from both inside the ring and on the watch glass outside the ring (Fig. 2o).
11. Move the ring (with the embryo) using coarse forceps from the watch glass to the 35 mm dish with albumen (Fig. 2p, q).
12. The vitelline membrane under the ring should bulge slightly upward. The embryo should be in the middle, at the highest point (Fig. 2r). The ring should naturally touch the bottom (remove some albumen if not) of the dish instead of floating on the albumen. Remove Pannett-Compton solution that has drained to the lowest point inside the ring. No albumen should leak inside the ring. Discard if so.
13. Put the lid on and put the petri dish inside a box moisturized with a wet towel. Incubate in a normal incubator (no need to have CO_2 or humidity control).
14. After a desired period of incubation, embryos are fixed and processed for further analysis. For example (Fig. 2s, t), embryos treated with a TGF-beta inhibitor from early-streak stage showed a reduction of Nanog expression in the central epiblast (Fig. 2t), whereas control-treated embryos exhibited normal expression (Fig. 2s).

3.3 Bead Graft and Subgerminal Cavity Injection

Epiblast-stage embryos can be experimentally manipulated using a number of techniques (*see* **Note 3**). Here we describe two of them, the bead graft technique and the subgerminal cavity injection technique.

3.3.1 Bead Graft

1. Embryos are prepared using the New culture method (Fig. 3a).
2. Prepare beads to be grafted. Beads come in different sizes. Pick the right bead size and pore size (for AG1 beads) suitable for your experiments. Bead preparation takes time, so carry out this step before or during New culture preparation. Aliquot 10–100 μl of beads to an Eppendorf tube. AG1 resins come as dry powdery beads, so aliquot an equivalent volume. These beads are relatively inexpensive and one bottle can last for

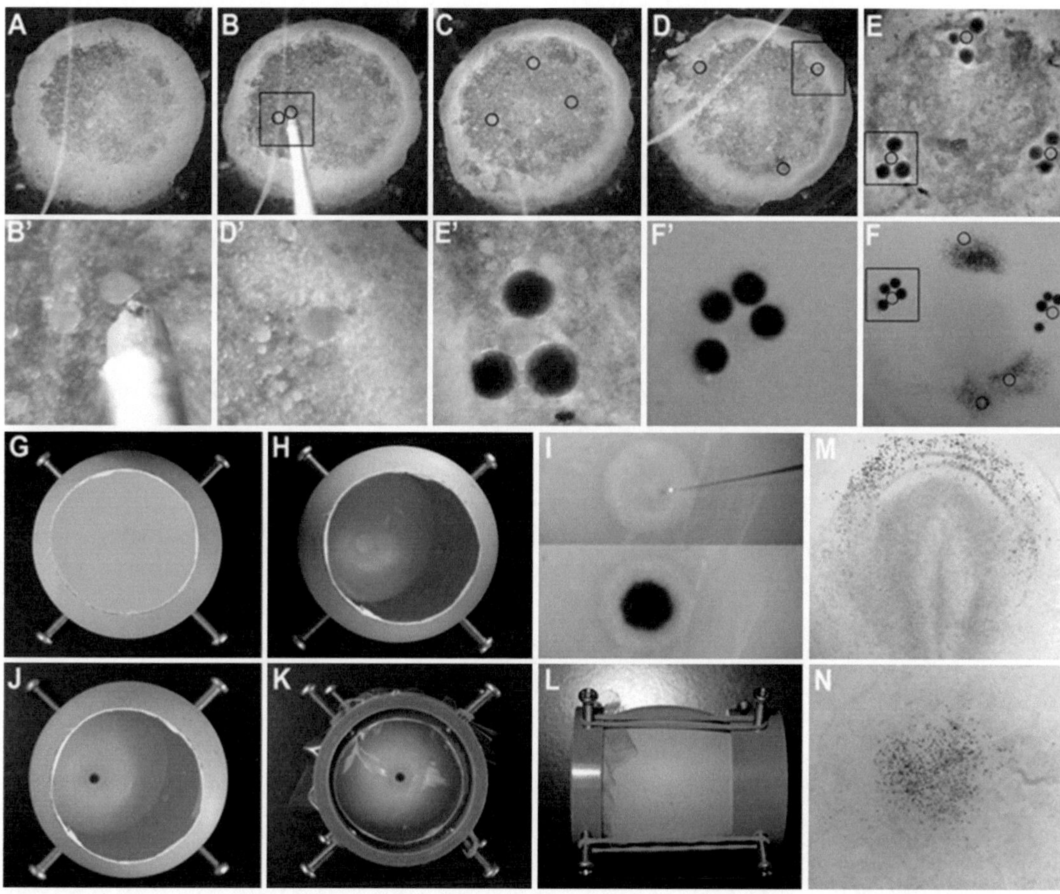

Fig. 3 Bead graft and subgerminal cavity injection. Embryos are prepared using the New culture setting. Growth factor-soaked beads are carefully placed on top of the embryo (**b, b′**) and positioned to the desired area using a sharp needle or metal tips (**c**, **d**, **d′**). Two types of beads can be combined (**e**, **e′**) with an example for such an experiment shown; induction of Brachyury by FGF can be inhibited when SU5402-soaked beads are placed in close proximity to FGF-loaded beads (**f**, **f′**). Unincubated eggs (embryo on top) are placed into prepared foster egg shells (**g**, **h**). Growth factor and/or small molecule solutions are injected into the subgerminal cavity using a fine pulled capillary and mouth pipetting (**i**, **j**). The injected egg is filled with albumen and sealed using plastic wrap, custom made rings, and rubber bands (**k**, **l**). Injected embryos are incubated horizontally and under rotation (90° every 90′). Examples of injected embryos are shown, with normal distribution of VASA positive PGCs in control-injected embryos (**m**), and abnormal number and location of PGCs after BMP injection (**n**)

many years. If the protein/small molecule to be loaded is dissolved in water-based solution as stock, wash the beads with PBS several times (each time a few minutes followed by short spins). If the stock is dissolved in DMSO, use DMSO to wash.

3. The stock solutions are normally kept as 5–10 μl aliquots, so only use equal or less volume of beads for loading. Mix beads and stock solution. Leave either on ice (for proteins) or RT (for small molecules) for at least a couple of hours.
4. Before use, wash beads 3× in PBS or Pannett-Compton solution. We wash by taking a couple of microliters out from the Eppendorf tube, and transferring successively into fresh wash solution three times.
5. Several (1–5) beads or bead combinations can be grafted on to each embryo. Beads can be added to the New culture after final assembly (Fig. 3b, b′) or at the step shown in Fig. 2m. Use fine forceps, a P20 pipette or any fine tips to add beads. Some beads may be sticky, add these beads as close to targeted locations as possible.
6. Use a 1 ml syringe needle or other fine metal tips to move beads to targeted areas (Fig. 3c, d, d′).
7. If a combination of beads is required, add a second type of beads (Fig. 3e, e′).
8. When all beads are in their targeted locations, remove excessive liquid from over the embryo and inside the ring using a capillary and mouth pipetting. Press beads down gently to make sure all beads are in contact with the embryo, and not floating on top of it.
9. Incubate. After intended period of incubation, make sure beads are still in place. If not, discard.
10. Fix the embryo and analyze the effect of the bead graft. An example is shown in Fig. 3f, in which three FGF-loaded beads induce Brachyury gene expression, and this induction is inhibited when both FGF- and SU5402-loaded beads are grafted.

3.3.2 Subgerminal Cavity Injection

1. Prepare foster egg shells (top part evenly cut with a circular saw) and large plastic rings as shown in Fig. 3g.
2. Transfer egg content from an unincubated egg to the foster egg shell (Fig. 3h).
3. Reposition the yolk with a spoon so that the embryo is on the top.
4. Use a fine pulled capillary and mouth pipetting, take up 5–10 μl of solution to be injected (growth factors or small molecules diluted in PBS or Pannett-Compton Solution, with a bit of fast green dye for visualization).

5. Inject 1–1.5 μl of solution per embryo into the subgerminal cavity (Fig. 3i, j). Do not inject above the epiblast or into the yolk. This is easy to achieve with only a little practice. The capillary opening should be large enough so that 1–1.5 μl can be injected in 2–3 s, but small enough so that no injected content leaks out afterwards.
6. Fill the injected egg with albumen (use a 10 ml syringe) to the very top. Cover the egg with a small piece of plastic wrap. Leave no trapped air bubbles. Cover the wrap with another plastic ring (Fig. 3k).
7. Tie together the top and bottom rings with rubber bands (Fig. 3k, l).
8. Incubate horizontally with 45–90° rotation every 90′.
9. After incubation, pour out the content, obtain the embryo as shown in Fig. 1 and analyze the effect of injected materials.
10. An example is shown in Fig. 3m, n. Embryo with control injection shows normal PGC specification (Vasa staining), whereas abnormal location and number of PGCs are seen in BMP-injected embryo.

3.4 RNA In Situ Hybridization

Whole-mount RNA in situ hybridization is the most basic technique in gene expression analysis. To achieve good results, follow basic precautions for RNA in situ analysis (*see* **Note 4**).

1. Make DIG or Fluorescein-labeled probes. This is an important step, but is the same for any type of RNA in situ analysis. So it is only briefly explained here. First, clone the region you intend to make the antisense probe of. We routinely use PCR method and pGEM-T-Easy as the host vector. Second, confirm the sequence and check the orientation by sequencing. Keep each construct as a stock. Third, either linearize the construct by digestion for antisense-directed transcription, or PCR the insert out using M13F/R, purify the DNA. Fourth, using the DIG or Fluorescein labeling kit, transcribe the antisense probe. Fifth, purify the probe. Keep it as a concentrated stock in a few milliliters of hybridization solution. Sixth, test each probe before experimental use.
2. Collect embryos as shown in Subheading 3.1. Fix them flat.
3. Remove the fixative completely and add 100 % methanol (with one rinse). Embryos can be stored this way for several weeks at −20 °C. Strong and clean probes may work with embryos kept this way for a couple of years.
4. Remove methanol, rehydrate in PBT by rinsing a couple of times and washing (nutating) for 30 min (Fig. 4a). Do not worry about yolk materials attached to the embryo. They will

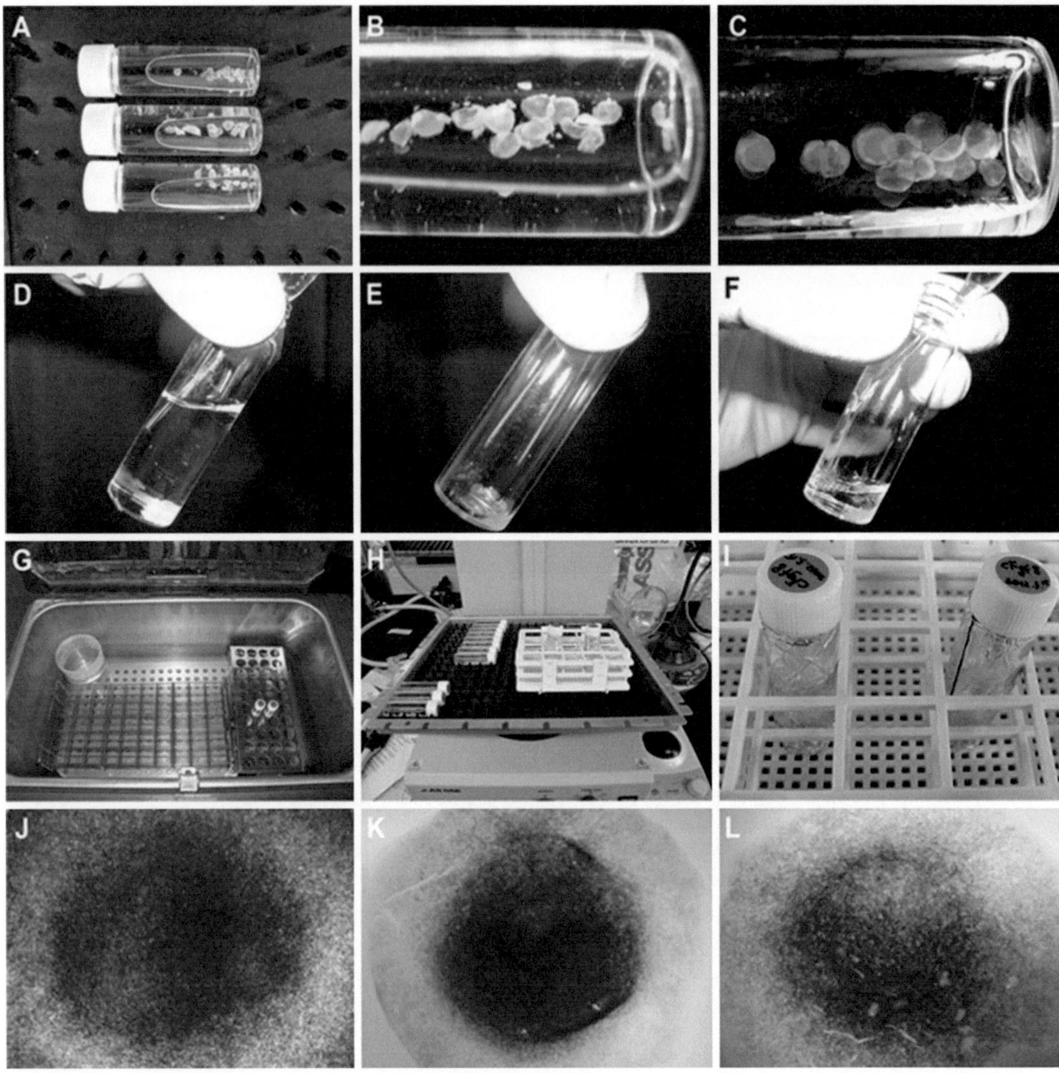

Fig. 4 RNA in situ hybridization. PFA and methanol fixed embryos (**a–c**) are pre-hybridized at 68 °C for several hours following Proteinase K treatment, post-fixation, and multiple PBT washes. Removal and addition of liquids should always be done from the side of the vial (**d–f**). Following pre-hybridization diluted DIG and/or Fluorescein-labeled probes are added to vials and incubated over night using a water bath at 68 °C (**g**). After multiple washing steps prior and post-incubation with an anti-DIG and/or anti-Fluorescein antibody solution, color development is initiated using a NTMT + BCIP + NBT solution (**h**, **i**). Following several hours up to several days of color development at RT or 4 °C, embryos are washed with TBST and used for further analysis. Representative in situ hybridization results of the pluripotency associated genes Nanog (**j**), Dnmt3B (**k**), and Lin28 (**l**)

come off during washes in subsequent steps (Fig. 4b) and embryos will eventually be free of yolk (Fig. 4c). When changing solutions in a vial, remove liquid from the side (do not touch or aspirate the embryos) and add fresh liquid to the side (do not drop directly onto the embryos) (Fig. 4d, e, f).

5. Digest the embryos with Proteinase K (Sigma) (10 μg/ml) for 15–30 min at RT. Leave the vials standing up, do not nutate. During and after this treatment, embryos are very fragile, but should not disintegrate (if so, adjust Proteinase K concentration or digestion time).
6. Remove Proteinase K very carefully, replace with postfix solution. Fix for 30 min to 1 h.
7. Remove postfix and wash with PBT (rinse 2× and wash 2× 10 min). Change into pre-hybridization solution.
8. Pre-hybridize in a water bath set at 68 °C (Fig. 4g) for at least 2 h. This step can be extended by several hours, or stopped by storing embryos in pre-hybridization solution at −20 °C for up to a couple of years.
9. Dilute labeled probes with pre-hybridization solution to desired dilutions (this should be tested out beforehand for each probe and diluted probes can be reused for many months). Diluted probe in pre-hybridization solution is called hybridization solution. Remove the pre-hybridization solution and add warmed-up hybridization solution. Hybridize overnight.
10. Remove the hybridization solution (and reuse) and wash with warmed-up pre-hybridization solution (3× rinses and 2× 30 min washes). Avoid contamination when working with multiple probes.
11. Change into 1:1 mix of pre-hybridization solution and TBST solution. Keep for another 30 min at 68 °C.
12. Wash in TBST (2× rinses and 3× 30 min washes) at RT.
13. Block in the blocking solution for 2 h at RT.
14. Incubate in the antibody solution (blocking solution + anti-DIG or anti-Fluorescein antibody (1:5,000 dilution)) overnight at 4 °C.
15. Wash in TBST (3× rinses and 3× 60 min washes) at RT.
16. Wash in NTMT for 10 min. Proceed to color development (NTMT + NBT + BCIP solution, protect from light) (Fig. 4h, i).
17. Monitor the color development in the first hour at RT. Decide on how long to continue at RT or move to 4 °C. If there is no signal in the first hour, the reaction can be left for several more hours at RT. If there are weak signals in the first hour, monitor closely every hour or so at RT or leave at 4 °C overnight. A good probe for a highly expressed gene will reach ideal signal/background ratio in 1–2 h. But for most moderately and specifically expressed genes, an ideal outcome is to have the best signal/background ratio within from several hours to overnight. Sometimes color reaction may take a few days. It is still good if there is no background or trapping signals.

18. To stop the color reaction, wash with TBST a few times (overnight if staining is strong). Change into fix solution and store at 4 °C for further analysis (photography, secondary staining, or sectioning).
19. Examples of in situ hybridization of pluripotency genes are shown in Fig. 4j (Nanog), Fig. 4k (Dnmt3B), and Fig. 4l (Lin28).

3.5 Primary Culture of Dissociated Epiblast Cells and Subculture of Epi-SCs

Refer to **Note 5** for a general description of avian epiblast-derived ES cells. Here we show how to make recombinant chicken LIF, prepare feeder layer, generate primary epiblast cell culture, and maintain Epi-SCs.

3.5.1 Production of Recombinant Chicken LIF

1. Total RNA is extracted from epiblast cells (*see* Subheading 3.1) with Trizol reagent and is used for reverse transcription PCR (RT-PCR) with the SuperScript™ III. Design and synthesize the appropriate oligonucleotide primers for amplification of chicken LIF. For cloning into pSecTag2A vector, modify the 5′-end sequence of the primers: forward primer 5′-GC*GCTAGC*C<u>ATG</u>AGGCTCATCCCC-3′ (italics, *Nhe*I restriction site; underline, start codon); reverse primer 5′-GC*GTCGAC*CGCGGGGCTGAGGTGAGG-3′ (italics, *Sal*I restriction site).
2. Prepare the PCR mixture (50 μl final volume) in an amplification tube: 1.0 μl template DNA (approx. 100 ng), 5.0 μl 2.5 mM dNTPs, 5.0 μl 10× Ex Taq buffer, 1.0 μl forward primer (10 pmol/μl), 1.0 μl reverse primer (10 pmol/μl), 0.5 μl Takara Ex Taq DNA polymerase (5 U/μl), and 36.5 μl H_2O. Using a thermal cycler, heat the samples to 94 °C for 5 min and then run 30 amplification cycles in the linear range of 30 s at 94 °C (denaturation), 30 s at 60 °C (annealing), and 1 min at 72 °C (polymerization). Finally, hold for 10 min at 72 °C as an extension step and then store at 4 °C.
3. Digest both pSecTag2A and PCR products with *Nhe*I and *Sal*I, and purify the DNAs using a commercially available kit. The myc-epitope in pSecTag2A is excised from the vector using restriction enzymes. Ligate the PCR products into the pSecTag2A vector using a commercially available kit. Transform the recombinant vector to TOP10F *E. coli* strain and purify the recombinant vector using EndoFree Plasmid Maxi Kit.
4. Grow up CHO cells in Ham-F12 supplemented 10 % FBS or CHCC-OU2 cells in DMEM supplemented with 10 % FBS in a 6-well culture plate. The seeded cells are 90–95 % confluent at the time of transfection. Transfect 2.0 μg/ml of the recombinant vector using Lipofectamine 2000 reagent.
5. After incubation for 24 h, remove the medium including the recombinant vector and add fresh medium supplemented with

10 % FBS and 250 μg/ml of Zeocin. Feed the cells with the selective medium every 3 days until foci can be identified. Pick and expand 20–40 foci to obtain recombinant chicken LIF producing cells. Check the producing cells by immunoblotting analysis using the culture supernatant and anti-His HRP antibody (supplied with the kit) (Fig. 5a). Grow up the cells in DMEM supplemented with 10 % FBS and collect the culture supernatant of 500 ml.

6. Recombinant chicken LIF is expressed as a fusion protein linked to a 6× Histidine tag and purified by affinity chromatography method using ProBond Resin. Add 10 ml of the washed resin to 500 ml of the culture supernatant, and aliquot the supernatant in 50 ml tubes. Use gentle agitation such as end-over-end rotation for 2 h at 4 °C. Pour the resin in a 2.5 cm × 10 cm column. Wash the column with 100 ml native wash buffer (supplied with the kit). Repeat washing three more times.
7. Elute the protein with native elution buffer (supplied with the kit). Collect 10 ml fractions and analyze with SDS-PAGE (Fig. 5b). Pool the major fractions containing the recombinant chicken LIF in a clean tube. Dialyze against cold PBS at 4 °C overnight. If necessary, concentrate to about 100 μg/ml using Amicon Ultra-15.
8. Aliquot the recombinant chicken LIF into 1.5 ml tubes, store at −80 °C until use. Using this method, 0.5–2 mg of purified recombinant chicken LIF can be obtained.

3.5.2 Preparation of Feeder Layer

1. Grow STO cells in DMEM supplemented with 10 % FBS until confluent.
2. Remove medium and add fresh medium supplemented with 10 % FBS and 10 μg/ml of mitomycin C.
3. After incubation for 2 h, remove the medium, and wash the culture dishes in PBS (4× rinses) at RT.
4. Add PBS with 0.025 % trypsin and 1 mM EGTA, and incubate for 5 min.
5. Resuspend the cells in a volume of DMEM supplemented with 10 % FBS and spin the cells in the DMEM for 5 min at 500 × *g* (3× rinses) at RT.
6. Plate the cells to gelatin-coated culture dishes at a concentration of 10,000–15,000 cells/cm^2 and culture at 37 °C with 5 % CO_2 for 24 h. The culture dishes are gelatinized with 0.1 % gelatin solution for a minimum of 30 min before use.
7. STO feeder layers should be prepared and used within 4 days.

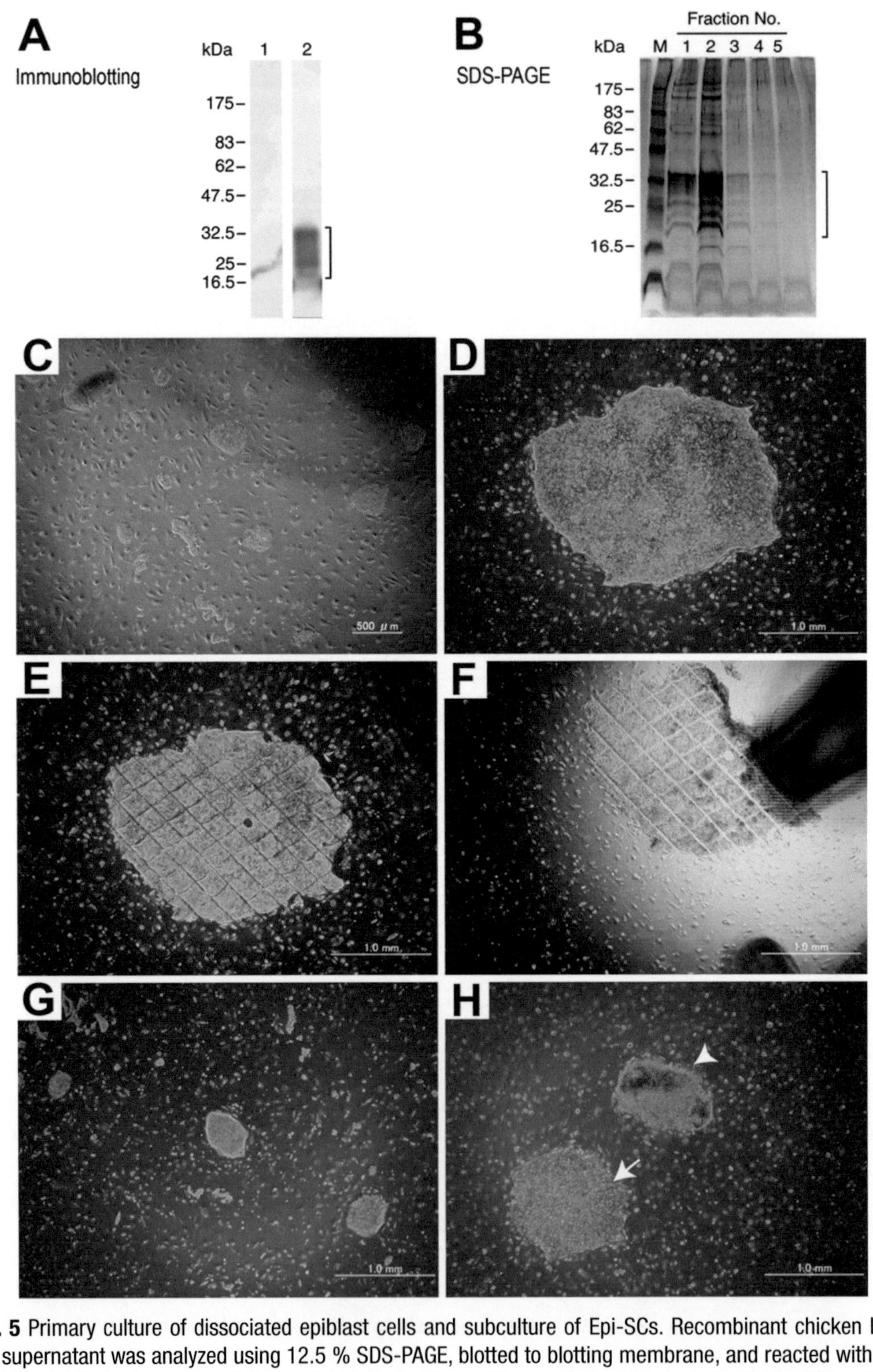

Fig. 5 Primary culture of dissociated epiblast cells and subculture of Epi-SCs. Recombinant chicken LIF in the supernatant was analyzed using 12.5 % SDS-PAGE, blotted to blotting membrane, and reacted with anti-His HRP antibody. *Lanes 1* and *2* were incubated with negative control antibody and anti-His HRP antibody, respectively (**a**). The fractions were analyzed using 15 % SDS-PAGE and the silver-staining method. Molecular

3.5.3 Primary Culture of Epiblast Cells

1. Warm ECC medium and KO-DMEM at 37 °C and keep them warm until ready for use.
2. Pick up dissected epiblast tissues (*see* Subheading 3.1 and Fig. 1m), and transfer them to a 1.5 ml siliconized tube.
3. Centrifuge at 120×*g* for 3 min, and then remove the supernatant.
4. Add 0.5 ml of ECC medium and gently suspend with a micropipette.
5. Remove the culture medium from STO feeder layer plated on a 35 mm culture dish, wash the dish with KO-DMEM, and add 2 ml of ECC medium.
6. Add 0.5 ml of the cell suspension to the culture dish and culture at 37 °C with 5 % CO_2 and 3 % O_2.
7. After 2 days add 0.5 ml of ECC medium. The cultured epiblast cells form small and round colonies (Fig. 5c).
8. Change one-half of the total medium every 2 days. After 5–6 days, the colonies grow to the size of approximately 1 mm in diameter (Fig. 5d).

3.5.4 Subculture and Maintenance of Epi-SCs

1. After 5–6 days, the colonies of cultured epiblast cells are ready for mechanical passaging by using a razor or a stem cell passaging tool. Do not cut up the colonies too finely at this step.
2. Warm ECC medium and KO-DMEM at 37 °C and keep them warm until ready for use.
3. Aspirate the medium from the epiblast cell cultured dish. Wash once with 2 ml of ECC medium. Replace with 1 ml of ECC medium.
4. Cut the colonies using a razor or a stem cell passaging tool under the stereomicroscope (Fig. 5e). Collect the segmented colonies with a micropipette (Fig. 5f) and break up cell clumps by gentle pipetting.
5. Seed these cells to new 35 mm dish(es) plated with feeder layer cells (typically at a 1:1 or 1:2 passaging ratio).
6. After 1 day, small colonies start to form (Fig. 5g). After 3 days from passaging, both healthy (arrow) and unhealthy

Fig. 5 (continued) size markers: *lane M*. The molecular mass of recombinant chicken LIF is 20–35 kDa (**b**). When dissociated epiblast cells are cultured with ECC medium, they form small and round colonies after 2 days (**c**). After 6 days, the colonies grow to the size of over 1 mm in diameter (**d**). For subculture, colonies are cut with a stem cell passaging tool (**e**). The segmented colonies are collected with a micropipette (**f**). After 1 day from passaging, small colonies start to form (**g**). After 3 days from passaging, both healthy (*arrow*) and unhealthy (*arrowhead*) large colonies start to form (**h**) (arrowhead) large colonies start to form (Fig. 5h). Remove the unhealthy colonies with a micropipette.

7. After 4–5 days from passaging, remove the unhealthy colonies and subculture only the healthy colonies. Change half of the total medium every 2 days.
8. Repeat **steps 3–7**. After about 3 weeks of primary culture, these cells can grow stably and can be considered to be chicken Epi-SCs.

3.6 Evaluation of Pluripotency in Epi-SCs

Currently, tools for evaluating the pluripotency of cultured chicken Epi-SCs are limited (*see* **Note 6**). This is due primarily to the lack of good antibodies for pluripotency markers and of suitable in vivo differentiation assays. Here we describe two basic evaluation protocols: RT-PCR analysis for Oct4 and Nanog, and immunostaining analysis for Nanog.

3.6.1 mRNA Expression Analysis of Oct4 and Nanog

1. Using a micropipette, collect the Epi-SC colonies from a 60 mm culture dish into a 1.5 ml Eppendorf tube.
2. Wash the cells in RNase-free PBS by mild centrifugation. Remove the supernatant carefully.
3. Total RNA is extracted from the Epi-SCs (Trizol method) and is used for reverse transcription PCR (RT-PCR) with SuperScript™ III. The primers for expression analysis of chicken Oct4, Nanog, and GAPDH (internal control) are shown in Fig. 6a.
4. Prepare the PCR mixture (50 μl final volume): 1.0 μl template DNA (approx. 100 ng), 5.0 μl 2.5 mM dNTPs, 5.0 μl 10× Ex Taq buffer, 1.0 μl forward primer (10 pmol/μl), 1.0 μl reverse primer (10 pmol/μl), 0.5 μl Takara Ex Taq DNA polymerase (5 U/μl), and 36.5 μl H_2O.
5. Using a thermal cycler, heat the samples to 94 °C for 5 min and then run optimal amplification cycles in the linear range of 30 s at 94 °C (denaturation), 30 s at optimal temperature (annealing, *see* Fig. 6a), and 1 min at 72 °C (polymerization). Finally, hold for 10 min at 72 °C as an extension step and then store at 4 °C.
6. Analyze the amplification products on a 1.5–2 % agarose gel (Fig. 6b).

3.6.2 Immunostaining of Nanog

1. Gelatinize chamber slides with 0.1 % gelatin solution for 1 day.
2. Plate mitomycin C-treated STO cells (*see* Subheading 3.5.2) to the slides at a concentration of 10,000–15,000 cells/cm^2 and culture at 37 °C with 5 % CO_2 for 24 h.
3. Aspirate the medium from the slides. Wash once with ECC medium and replace with ECC medium.

A Primers for exoression analysis using RT-PCR

Genns		Sequences	Annealing (C°)	Cycles
Nanog				
	forward	TGACCACAGAGCAGAAAACG	58	24
	reverse	CAGCCATGAACGGATACAGG		
Oct4				
	forward	GGGGCTGGCTCTGGGCACGCTCTAT	60	22
	reverse	CCTGGGGACTGGGCTTCACACATTT		
GAPDH				
	forward	GCACGCCATCACTATCTTCCAG	58	22
	reverse	CGGCAGGTCAGGTCAACAACAG		

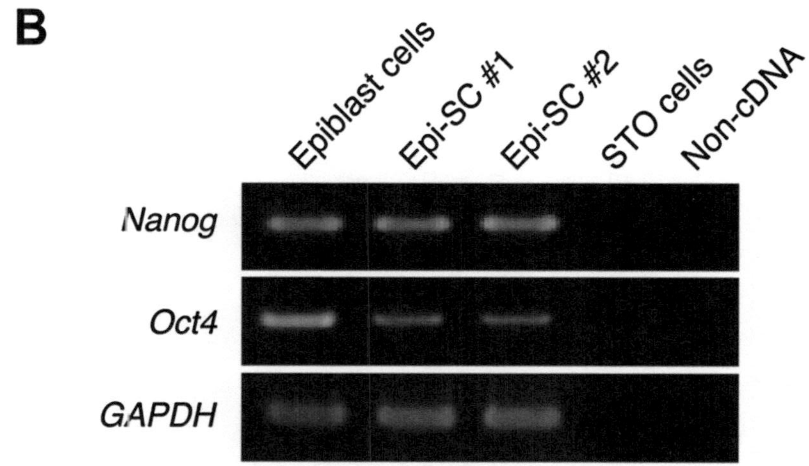

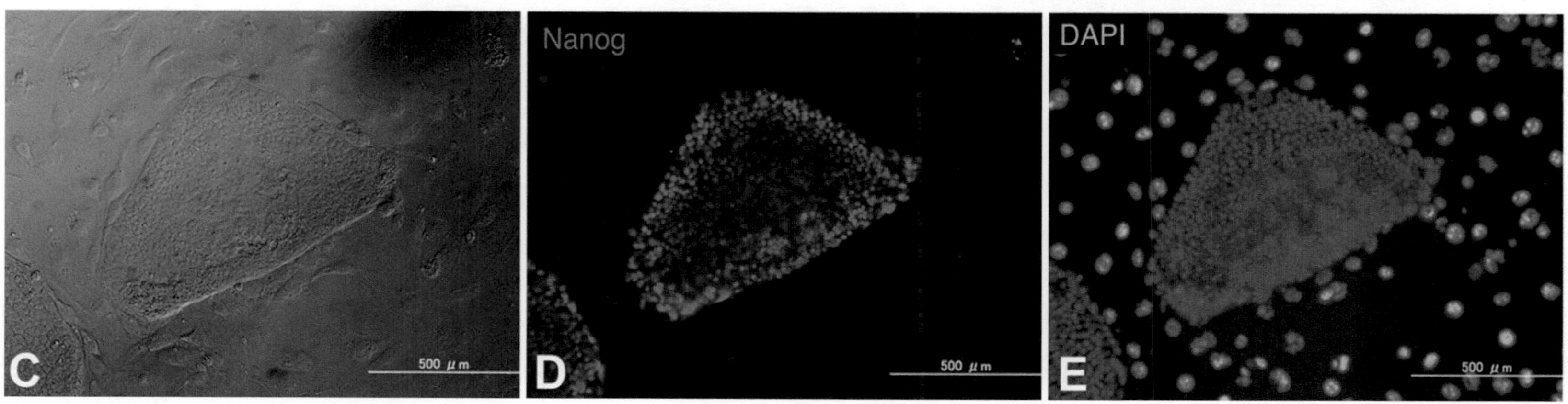

Fig. 6 Evaluation of pluripotency in Epi-SCs. Primer information for expression analysis of Nanog and Oct4 is shown in (**a**). RNAs are isolated from Epi-SCs successively subcultured for 30 days (#1 in **b**) and for 50 days (#2 in **b**). RNAs isolated from uncultured epiblast cells (positive control) and STO cells (negative control) are also examined with the same primers. Chicken GAPDH gene is used as an internal control. In immuno-analysis, Nanog is seen to be expressed in the nucleus of most cells of an Epi-SC colony (**c**–**e**)

4. Seed the segmented colonies of Epi-SCs to the slides and culture at 37 °C with 5 % CO_2 and 3 % O_2. After 2 or 3 days, aspirate the medium from the slides and wash with cold PBS (3× rinses).
5. Add 4 % PFA and incubate for 30 min at RT. Wash in 10 mM glycine–PBS (3× rinses) and rewash in cold PBS (2× rinses).
6. Add 0.1 % Triton-X-100–PBS and incubate for 5 min at RT. Wash in cold PBS (3× rinses).
7. Block in 3 % BSA–PBS for 15 min at RT. Aspirate the blocking solution and add ×100 of anti-chicken Nanog polyclonal rabbit antibody diluted in 1 % BSA–PBS for 1 h at RT. Wash in 0.1 % BSA–PBS (6× rinses).
8. Add ×100 of Alexa Fluor 594-conjugated anti-rabbit IgG antibody and incubate for 1 h at RT. Wash in 0.1 % BSA–PBS (6× rinses).
9. Mount the slide with Vectashield mounting media. The sample is examined using fluorescence microscopy (Fig. 6c–e).

3.7 General Remarks

The chick model is an essential complement to the mouse/human models in epiblast biology. The mammals and reptiles (including birds) are two main branches of the amniotic vertebrates, and development of primitive mammals has a lot in common with that of the reptiles. Eutherian mammals (humans, mice, etc.) have many derived features in their early development which are difficult to understand without the knowledge of comparative amniote embryology. Studies using the chick, the main nonmammalian model organism, can therefore offer essential insights in understanding the formation, differentiation, and pluripotency maintenance of the epiblast and epiblast stem cells.

4 Notes

1. Fertilized and freshly laid eggs purchased from poultry farms can be stored at 16 °C for up to 1 week. Eggs are either warmed up briefly (30′) for the unincubated stage or incubated at 38.5 °C for up to 18 h to reach desired stages. Depending on the weather and flock conditions, developmental stage of embryos from unincubated eggs varies, and this needs to be taken into consideration when designing an experiment. Epiblast-stage embryos are fragile and easily damaged by metal tools or at the liquid/air interface. Keep embryos submerged in liquid and do not grab them directly with forceps. A basic understanding of how to stage and operate on early-streak stage embryos is necessary [6].

2. New culture was invented by Denis New [7] with several modifications made afterwards [8]. It has been the most powerful ex ovo culture method for pre-streak and early-streak stage chicken embryos. Using this culture method, young embryos can be experimented upon using a variety of techniques (tissue graft, bead graft, DNA electroporation, time-lapse imaging, and whole embryo treatment with small molecules). Although it is tedious to set up and takes practice to master, the New culture and its versatile applications are indispensible for avian epiblast-related studies.
3. To study a specific gene or pathway during early avian development, localized treatment is often preferred. The New culture method (Subheading 3.2) can be modified for such purpose. Localized effects can be achieved through bead graft or targeted DNA electroporation. For protein absorption, Heparin-coated agarose/acrylic beads or Affigel blue beads are commonly used. We have also successfully used hydrogel to embed and release proteins locally. For absorption of small molecules, AG1 ion exchange beads are used. A great variety of small molecules (e.g., agonists or antagonists of kinases, receptors, or membrane channels) are currently available. But proteins are limited to secreted growth factors. They are often expensive to acquire from commercial sources and in general their efficacy in such experimental settings has not been tested. This limitation is circumvented by DNA electroporation. DNA constructs expressing any gene of interest can be electroporated and their effect analyzed locally in cells expressing the gene of interest, either through a protein tag, 2A-peptide-mediated co-expression of a marker gene, or co-electroporation of a marker gene-encoding construct. The drawback of electroporation, especially for early-stage embryos, is that there is a minimum of 2 h (for GFP) needed to transcribe and translate introduced genes. Any effect from an electroporated gene may not be evident until at least 4–6 h culture after electroporation. In addition to the New culture-based methods, we have also tested a method for in ovo analysis of the epiblast-stage embryos. We call it the subgerminal cavity method. The subgerminal cavity is the space between the yolk cell membrane and the epiblast/hypoblast. Its lateral margin is the germ wall where the yolk cell and deep layer cells of the germ wall adhere to each other. In unincubated eggs, epiblast cells are directly exposed to the subgerminal cavity. Growth factors and/or small molecules can be injected into this space, and their influence on epiblast cell differentiation can be analyzed after a short period of incubation. The advantage of this method is that it is in ovo, easy to learn and quick to perform (compared to the New culture) and its effect

on epiblast or embryonic development is rapid (compared to the electroporation method).

4. There are a few protocols for RNA in situ hybridization of chicken embryos. The main differences are in the ingredients for the hybridization solution and the hybridization temperature. We follow the protocol of the Stern lab [9]. This protocol has worked well for us. We have also used the same protocol successfully for reptilian and other avian embryos. Pre-gastrulation and early gastrulation-stage embryos are fragile, so throughout the entire in situ process (solution changes, washing, etc.), care should be taken to avoid damage to the embryos. The "tissues" at these stages are only a few cell-layers thick, so there is minimal worry of tissue trapping or trapping-related background staining. Key to success is the quality of DIG- or Fluorescein-labeled probes and the washes after hybridization and antibody incubation. Routine precautions for RNA-related work (baking glassware and metal tools and using DEPC-treated water) should be taken when preparing stock solutions.

5. Chicken epiblast cells derived from a stage X (EGK) embryo have the capacity to contribute to all three germ layers and germ cells when transplanted into a recipient embryo of the same developmental stage [10, 11]. Thus, attempts have been made to establish chicken embryonic stem (ES) cells by in vitro culture of early epiblast cells [5, 12–14]. All of the established ES cells are of the epiblast origin, and we describe them as epiblast stem cells (Epi-SC). The pluripotency and growth of cultured epiblast cells can be maintained by the addition of chicken leukemia inhibitory factor (LIF) [15] or of conditioned medium from buffalo rat liver cells [13] to the culture medium. We follow the protocol of epiblast culture using chicken LIF. Recombinant chicken LIF protein [5, 16] is obtained from recombinant CHO-K1 cells (ATCC #CCL-61) or CHCC-OU2 cells (supplied by the United States Department of Agriculture [17]) transfected with the chicken LIF gene. The epiblast cells or Epi-SCs are plated onto a layer of mitotically inactivated Sandoz inbred mouse-derived thioguanine-resistant and ouabain-resistant (STO) cell (ATCC #CRL-1503) feeders. Chicken epiblast stem cells in culture are similar in morphology to primate ES cells and consist of circular colonies with clear contour and each individual cell visible.

6. Mouse and human ES cells express a number of pluripotency markers (Oct4, Nanog, Sox2, Rex1, Utf, etc.). In chickens, however, only Oct4 and Nanog have been reported [3, 5]. These factors are required for the maintenance of pluripotency and self-renewal of chicken ES cells [16]. Our unpublished results indicate that several other pluripotency markers, such as

Dnmt3B and Lin28, can also serve as good markers for avian Epi-SCs. The list will undoubtedly increase as more whole-genome analyses are being performed for early-stage chicken embryos. Pluripotency of cultured chicken Epi-SCs can be evaluated by assaying the expression level of these genes using RT-PCR or immunostaining. For RT-PCR analysis, the GenBank accession numbers for chicken Oct4 and Nanog are DQ867024 and DQ867025, respectively. For immunostaining, unfortunately, specific antibodies against these proteins are not commercially available, and the commercially supplied antibodies against their mammalian orthologues do not show satisfactory cross-reactivity.

References

1. Eyal-Giladi H, Kochav S (1976) From cleavage to primitive streak formation: a complementary normal table and a new look at the first stages of the development of the chick. I. General morphology. Dev Biol 49:321–337
2. Hamburger V, Hamilton HL (1992) A series of normal stages in the development of the chick embryo. 1951. Dev Dyn 195:231–272
3. Lavial F, Acloque H, Bertocchini F et al (2007) The Oct4 homologue PouV and Nanog regulate pluripotency in chicken embryonic stem cells. Development 134:3549–3563
4. Shin M, Alev C, Wu Y et al (2011) Activin/TGF-beta signaling regulates Nanog expression in the epiblast during gastrulation. Mech Dev 128:268–278
5. Nakano M, Arisawa K, Yokoyama S et al (2011) Characteristics of novel chicken embryonic stem cells established using chicken leukemia inhibitory factor. J Poult Sci 48:64–72
6. Streit A, Stern CD (2008) Operations on primitive streak stage avian embryos. Methods Cell Biol 87:3–17
7. New DAT (1955) A new technique for the cultivation of the chick embryo in vitro. J Embryol Exp Morph 3:320–331
8. Stern CD, Ireland GW (1981) An integrated experimental study of endoderm formation in avian embryos. Anat Embryol (Berl) 163: 245–263
9. Stern CD (1998) Detection of multiple gene products simultaneously by in situ hybridization and immunohistochemistry in whole mounts of avian embryos. Curr Top Dev Biol 36:223–243
10. Petitte JN, Clark ME, Liu G et al (1990) Production of somatic and germline chimeras in the chicken by transfer of early blastodermal cells. Development 108:185–189
11. Kagami H, Clark ME, Verrinder Gibbins AM et al (1995) Sexual differentiation of chimeric chickens containing ZZ and ZW cells in the germline. Mol Reprod Dev 42:379–387
12. Pain B, Clark ME, Shen M et al (1996) Long-term in vitro culture and characterisation of avian embryonic stem cells with multiple morphogenetic potentialities. Development 122: 2339–2348
13. van de Lavoir MC, Mather-Love C, Leighton P et al (2006) High-grade transgenic somatic chimeras from chicken embryonic stem cells. Mech Dev 123:31–41
14. Lavial F, Pain B (2010) Chicken embryonic stem cells as a non-mammalian embryonic stem cell model. Dev Growth Differ 52:101–114
15. Horiuchi H, Tategaki A, Yamashita Y et al (2004) Chicken leukemia inhibitory factor maintains chicken embryonic stem cells in the undifferentiated state. J Biol Chem 279: 24514–24520
16. Yamashita Y, Tategaki A, Ogawa M et al (2006) Effect of novel monoclonal antibodies on LIF-induced signaling in chicken blastodermal cells. Dev Comp Immunol 30:513–522
17. Lillehoj HS, Choi KD (1998) Recombinant chicken interferon-gamma-mediated inhibition of Eimeria tenella development in vitro and reduction of oocyst production and body weight loss following Eimeria acervulina challenge infection. Avian Dis 42:307–314

Chapter 13

Stepwise Differentiation from Naïve State Pluripotent Stem Cells to Functional Primordial Germ Cells Through an Epiblast-Like State

Katsuhiko Hayashi and Mitinori Saitou

Abstract

A group of pluripotent cells appearing during mammalian embryogenesis is the source for all the cell lineages that compose the embryo proper. In mice, pluripotent cells are first established in the inner cell mass (ICM) of the preimplantation blastocyst. After implantation, the ICM soon transforms into a cup-shaped epithelium, called the postimplantation epiblast. The two types of pluripotent cells, the ICM and postimplantation epiblast cells, are distinct, based on the differences in their gene expression profiles, epigenetic status, and differentiation capacity. During gastrulation, some of the postimplantation epiblast cells adjacent to the extraembryonic ectoderm are specified as primordial germ cells (PGCs), precursors of the germ cell lineage, in response to bone morphogenetic protein 4 (BMP4). Recently, we succeeded in reconstituting epiblast differentiation and PGC specification in vitro using pluripotent stem cells. Here, we describe the culture method of a stepwise differentiation from pluripotent stem cells to functional PGCs.

Key words Embryonic stem cells, Inner cell mass, Postimplantation epiblast, Primordial germ cells, BMP4

1 Introduction

During mammalian embryogenesis, cleaved blastomeres are sorted into two types of cell lineage, the inner cell mass (ICM) and trophectoderm; the former is a group of pluripotent cells forming the embryo proper and the latter is a precursor of the placenta. After implantation, the ICM is further sorted into the postimplantation epiblast and primitive endoderm. The postimplantation epiblast maintains pluripotency and, upon gastrulation, differentiates into various somatic cells and primordial gem cells (PGCs). PGCs, precursors of oocytes and sperm, are segregated from the somatic cell lineage in response to BMP4 secreted from the adjacent extraembryonic ectoderm [1]. PGCs possess a specific gene expression program that inhibits the somatic cell program and reacquires

Ramiro Alberio (ed.), *Epiblast Stem Cells: Methods and Protocols*, Methods in Molecular Biology, vol. 1074,
DOI 10.1007/978-1-62703-628-3_13, © Springer Science+Business Media New York 2013

potential pluripotency [2, 3]. Controlled by the specific gene expression program, PGCs reorganize the epigenetic status of the genome (epigenetic reprogramming) at a genome-wide level [4–6]. It is of particular importance to know in detail how epigenetic reprogramming is accomplished in PGCs, as a failure in epigenetic reprogramming may cause developmental disorder and infertility. However, only a limited number of nascent PGCs are available in vivo, limiting research.

To compensate for the limited number of nascent PGCs in vivo, it could be an option to reconstitute PGC specification processes in vitro by using pluripotent stem cells derived from either the ICM or postimplantation epiblast. Indeed, such attempts have been made using embryonic stem cells (ESCs) and epiblast stem cells (EpiSCs), pluripotent stem cells indefinitely proliferating in vitro that are derived from the ICM and postimplantation epiblast, respectively [7–10]. Despite the intensive efforts, a culture system that rigorously reconstitutes PGC specification and produces a robust number of nascent PGCs has been not established [11–14]. This is partially because the attempts at such systems have not paid enough attention to reconstitute the state of the postimplantation epiblast in vivo that produces PGCs in response to BMP4. We recently succeeded in producing a novel type of cells harboring the postimplantation epiblast status, called epiblast-like cells (EpiLCs), by inducing the transient differentiation of ESCs under a defined condition [15]. The EpiLCs differentiate robustly into PGC-like cells (PGCLCs) in response to BMP4 [15]. The characteristics of PGCLCs are highly similar, if not identical, to those of PGCs in vivo, based on the criteria of gene expression, epigenetic status, and potential to differentiate into fertile sperm. Here we describe the materials and methods used in establishing a culture system for the production of EpiLCs and PGCLCs.

2 Materials

1. Insulin: Prepare a 25 mg/ml stock solution dissolved with sterile 0.01 M HCl overnight at 4°C. Store at −30 °C.
2. Apo-transferrin: Prepare a 100 mg/ml stock solution dissolved with distilled water overnight at 4 °C. Store at −30 °C.
3. Progesterone: Prepare a 0.6 mg/ml stock solution dissolved with ethanol. Filter through a 0.22-μm syringe filter and store at −30 °C.
4. Putrescine: Prepare a 160 mg/ml stock solution dissolved with distilled water. Filter through a 0.22-μm syringe filter and store at −30 °C.
5. Sodium selenite: Prepare a 3 mM stock solution dissolved with distilled water. Filter through a 0.22-μm syringe filter and store at −30 °C.

6. Bovine serum albumin (BSA) fraction V, 7.5 % solution. Store at 4 °C.
7. Dulbecco's modified Eagle's medium nutrient mixture F-12 Ham (DMEM/F12). Store at 4 °C. Note that DMEM/F12 with HEPES may cause nonspecific Blimp1 expression.
8. GMEM. Store at 4 °C.
9. Neurolbasal™ medium (Invitrogen). Store at 4 °C.
10. DMEM. Store at 4 °C.
11. PBS. Store at 4 °C.
12. B27 supplement minus vitamin A. Store at −30 °C.
13. L-Glutamine 200 mM. Make 1 and 5 ml aliquots and store at −80 °C.
14. MEM Nonessential amino acid (NEAA). Make 1 and 5 ml aliquots and store at −80 °C.
15. Sodium pyruvate solution 100 mM. Make 1 and 5 ml aliquots and store at −80 °C.
16. Penicillin/Streptomycin. Make 1 and 5 ml aliquots and store at −80 °C.
17. β-Mercaptoethanol, 50 mM. Store at 4 °C.
18. Knockout Serum Replacement. Make 1 and 5 ml aliquots and store at −80 °C (*see* **Note 1**).
19. PD0325901 (Stemgent): Prepare a 10 mM stock solution dissolved with dimethyl sulfoxide (DMSO). Make a 20 μl aliquot and store at −80 °C.
20. CHIR99021 (Biovision): Prepare a 30 mM stock solution dissolved with DMSO. Make a 20 μl aliquot and store at −80 °C.
21. ESGRO (Millipore). Store at 4 °C.
22. Recombinant human Activin A: Prepare a 50 μg/ml stock solution dissolved with distilled water. Make a 20 μl aliquot and store at −30 °C.
23. Basic FGF (bFGF): Prepare a 10 μg/ml stock solution dissolved with 10 mM Tris, pH 7.6, containing 0.1 % BSA. Make a 50 μl aliquot and store at −30 °C.
24. Recombinant human BMP4 (R&D Systems): Prepare a 50 μg/ml stock solution dissolved with 4 mM HCl containing 0.1 % BSA. Make a 100 μl aliquot and store at −80 °C.
25. Recombinant human BMP8a (R&D Systems): Prepare a 50 μg/ml stock solution dissolved with 4 mM HCl containing 0.1 % BSA. Make a 100 μl aliquot and store at −80 °C (*see* **Note 2**).
26. Stem Cell Factor (SCF; R&D Systems): Prepare a 50 μg/ml stock solution dissolved with PBS containing 0.1 % BSA. Make a 20 μl aliquot and store at −80 °C.

27. Recombinant mouse EGF (R&D Systems): Prepare a 500 μg/ml stock solution dissolved with PBS containing 0.1 % BSA. Make a 20 μl aliquot and store at −80 °C.
28. Poly-L-ornithine: Prepare a 0.01 % stock solution dissolved with distilled water. Store at 4 °C.
29. Laminin, 1 mg/ml (BD Biosciences). Thaw and make a 20 μl aliquot on ice. Store at −30 °C. Thawed aliquots can be stored at 4 °C for up to 1 month. To prepare a working solution, dilute the stock solution to 10 ng/ml with PBS (*see* **Note 3**). Prepare freshly before use.
30. Human plasma fibronectin, 1 mg/ml (Millipore). Store at 4 °C.
31. TrypLE Express (Invitrogen). Store at 4 °C.
32. Trypsin–EDTA, 0.5 %: For a working solution, dilute ten times to obtain 0.05 % Trypsin–EDTA. Store at 4 °C.
33. 6-well multiwell plate.
34. 12-well multiwell plate.
35. Lipidure-Coat Plate A-U96 (NOF Corporation).
36. 15 ml conical centrifuge tube.
37. PE Anti-mouse/rat CD61 (BioLegend; catalogue no. 104307).
38. Anti-human/mouse SSEA1 Alexa Fluor647 (eBioscience; catalogue no. 51-8813).
39. Single-use filters, 0.22 μl.

2.1 Culture Media

1. 2i+LIF: The 2i+LIF medium was composed of N2B27 medium with PD0325901, CHIR99021, and LIF. N2B27 was prepared as described in a previous report [16] with slight modifications. For DMEM/F12+N2, mix 494 ml DMEM/F12 with 5 ml N2, 0.5 ml insulin stock solution, 0.5 ml apotransferrin stock solution, 0.33 ml 7.5 % BSA solution, 16.5 μl progesterone stock solution, 50 μl putrescine stock solution, and 5 μl sodium selenite stock solution. For Neurobasal+B27, mix 480 ml Neurobasal with 10 ml B27, 5 ml penicillin/streptomycin, and 5 ml L-glutamine stock solution. For 1 L of N2B27 medium, mix 500 ml DMEM/F12+N2 with 500 ml Neurolbasal+B27, and add 1.8 ml β-mercaptoethanol stock solution. Make a 40 ml aliquot of N2B27 and store at −80 °C. For 2i+LIF, mix 40 ml N2B27 with 4 μl CHIR99021 stock solution, 1.6 μl PD0325901 stock solution and 4 μl ESGRO. Store at 4 °C for up to 2 weeks.
2. EpiLC differentiation medium: Mix 5 ml N2B27 with 2 μl Activin A stock solution, 6 μl bFGF stock solution, and 50 μl KSR. Prepare freshly before use.

3. PGCLC differentiation medium: The PGCLC differentiation medium was composed of GK15 with 500 ng/ml BMP4, 500 ng/ml BMP8a, 100 ng/ml SCF, 1,000 U/ml LIF, and 50 ng/ml EGF. For GK15, mix 8.082 ml GMEM with 1.5 ml KSR, 100 μl nonessential amino acid, 100 μl sodium pyruvate, 100 μl L-glutamine, 100 μl penicillin/streptomycin, and 18 μl β-Mercaptoethanol. For 10 ml PGCLC differentiation medium, mix 9.778 ml GK15 with 100 μl BMP4, 100 μl BMP8a, 20 μl SCF, 1 μl LIF, and 1 μl EGF. Prepare freshly before use.
4. TrypLE-wash medium: Mix BSA into DMEM:F12 at a final concentration of 0.1 % BSA.

2.2 ES Cells

This protocol employs Blimp1-mVenus and Stella-ECFP (BVSC) reporter ES cells that monitor PGCLC differentiation [17], as well as non-reporter ES cells. Although both types of ES cells are useful for this method, the reporter ES cells are more suitable, since they facilitate detection of PGCLCs at an earlier stage [15].

3 Methods

This protocol is summarized in Fig. 1.

3.1 Culture of ES Cells

ES cells are cultured in 2i+LIF medium (see above) on a 6-well culture plate coated with poly-L-ornithine and laminin. To coat the culture dish, add 1 ml of 0.01 % ornithine solution into one well of a 6-well culture dish, and incubate for at least 1 h at room temperature. After washing the well twice with PBS, add 1 ml of 10 ng/ml laminin into the well (*see* **Note 3**) and incubate for at least 1 h at 37 °C. To dissociate ES cells when passaging, aspirate the 2i+LIF medium, pipet 1 ml of TrypLE Express and incubate for 4 min at 37 °C. After the incubation, add 4 ml of TrypLE-wash medium,

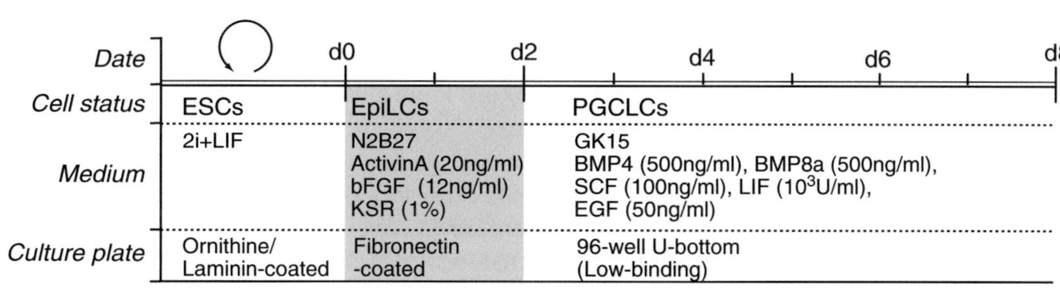

Fig. 1 Summary of procedure for EpiLC and PGCLC derivation from ES cells. ES cells (ESCs) are maintained in 2i+LIF medium on a culture plate coated with poly-L-ornithine and laminin. For EpiLC differentiation, ES cells are transferred into EpiLC differentiation medium and cultured on a fibronectin-coated culture plate for 2 days. After EpiLC differentiation, PGCLCs are induced in PGC differentiation medium

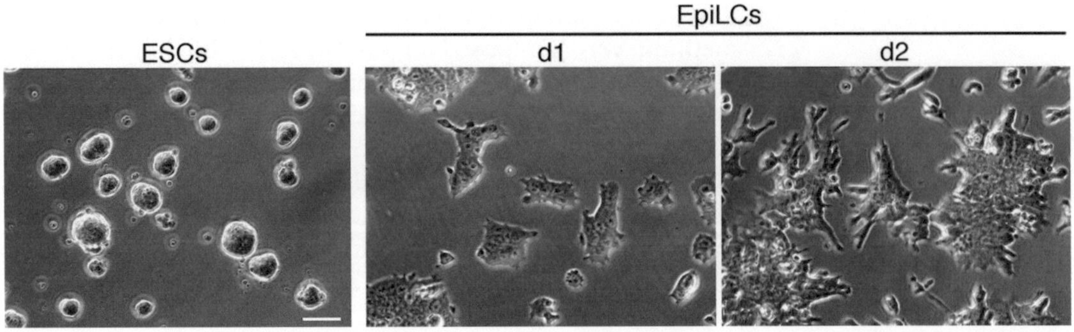

Fig. 2 EpiLC induction from ES cells. Shown are bright-field images ESCs and EpiLCs at day 1 (d1) and day 2 (d2) of culture. Scale bar, 50 μm. Note that morphology of ES cells appear flat in during EpiLC differentiation

pipette several times, and count the number of cells. Transfer a cell suspension containing 2.5×10^5 cells to a 15 ml conical tube (*see* **Note 4**), centrifuge at $220 \times g$, remove the supernatant, and resuspend with 2i+LIF. Place the cell suspension on an ornithine- and laminin-coated culture plate.

3.2 EpiLC Differentiation

For EpiLC differentiation, prepare a 12-well culture plate coated with human plasma fibronectin as follows: dilute 10 μl fibronectin with 600 μl PBS to obtain 16.6 μl/ml fibronectin solution, pipet the solution into one well of a 12-well plate, incubate for at least 1 h at 37 °C and then aspirate the fibronectin solution just before placing the ES cells. Dissociate the ES cells as described above in Subheading 3.1. Resuspend 1×10^5 cells with 1 ml EpiLC differentiation medium and place on a fibronectin-coated culture plate (*see* **Note 5**). Change the medium at day 1 of culture. 1×10^5 ES cells usually yield 6×10^5 to 1×10^6 EpiLCs at day 2 of culture. A representative result of EpiLC differentiation is shown in Fig. 2.

3.3 PGCLC Differentiation

To dissociate EpiLCs at day 2 of culture (*see* **Note 6**), aspirate EpiLC culture medium, wash with PBS, incubate with TrypLE express for 2 min at room temperature, add TrypLE-wash medium, and collect the cells by centrifugation. Resuspend EpiLCs in PGCLC differentiation medium and adjust the concentration of the cells to 1×10^4 cells/ml. Pipet 100 μl of the cell suspension into a single well of a Lipidure-Coat Plate (*see* **Note 7**). The cells in the well of the plate will form a small aggregate (Fig. 3a). If BVSC ES cells are used, BV and SC reporters will be detectable at day 1 or 2 and day 3 or 4 of culture, respectively. If non-reporter ES cells are used, SSEA1- and CD61-positive cells will be distinguishable at day 4 of culture.

3.4 FACS Sorting of PGCLCs

PGCLCs can be isolated by fluorescence-activated cell sorting (FACS). Collect aggregation containing PGCLCs by using either a glass capillary or pipette with a plastic tip, wash with PBS,

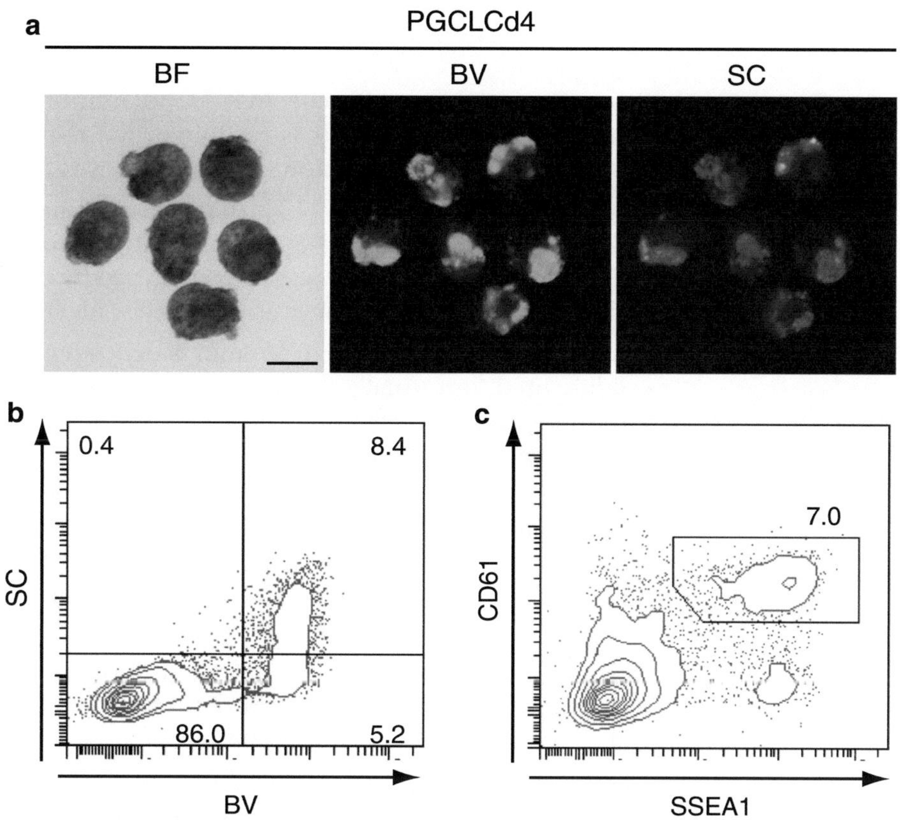

Fig. 3 Images of PGCLCs (BVSC) and the FACS pattern. (**a**) Shown are bright-field images and fluorescence images of PGCLCs at day 4 of PGCLC induction. Cells clearly expressing BV and SC are observed in the aggregates. Scale bar, 200 μm. (**b**, **c**) FACS analyses show percentages of BV- and SC-positive PGCLCs from the BVSC reporter ES cells (**b**) and of CD61- and SSEA1-positive PGCLCs from the non-reporter ES cells (**c**). PGCLCs at day 4 of PGCLC induction were used for the analyses

and dissociate the aggregations by incubation with 0.05 % Trypsin–EDTA at 37 °C. The incubation time depends on the stage of PGCLCs, being 6 min for PGCLCs at day 2 of culture, 8 min for PGCLCs at day 4 of culture, and 10 min for PGCLCs at day 6 of culture. Add an equal volume of DMEM containing 10 % fetal bovine serum for quenching trypsin–EDTA, and pipette several times to obtain single suspension. In the case of BVSC ES cells, PGCLCs can be immediately purified by FACS (Fig. 3b). Otherwise, the single cell suspension is incubated for 15 min on ice with PE anti-CD61 diluted at 1:200 and anti-SSEA1 Alexa Fluor647 diluted at 1:20. After washing with PBS containing 0.1 % BSA, cells resuspended with an appropriate volume of PBS containing 0.1 % BSA are sorted by FACS. Collect CD61- and SSEA1-positive cells (Fig. 3c), which are nearly identical to BV-positive cells [15].

4 Notes

1. Practically speaking, efficient PGCLC derivation sometime depends on the KSR lot. It is recommended that a KSR lot optimal for PGCLC derivation is chosen in advance.
2. After publishing our paper [15], the supplier announced that the name of BMP8b (R&D Systems; catalogue no. 1073-BP-010) was changed to BMP8a. The protein products of BMP8a and BMP8b share 98 % amino acid sequence identity.
3. The optimal concentration of laminin is dependent on the ES cell line used, but usually ranges from 10 ng/ml to 1 μg/ml. Find the optimal concentration by culturing ES cells on various concentrations of laminin.
4. The concentration of cells used is very important, as overgrowth immediately causes cell death, and on the other hand, a lower concentration delays cell proliferation. The optimal number of cells spread on the plate will sometimes depend on the particular ES cell line used.
5. The number of ES cells used for EpiLC differentiation is important. Using less than half of the total ES cells ($<5 \times 10^4$) will cause cell death at day 1 of culture.
6. The optimal timing for PGCLC differentiation is also sometimes dependent on the ES cell line used. It is recommended that PGCLC differentiation is induced in EpiLCs at various culture points, e.g., at 36, 42, and 48 h of culture.
7. EpiLCs appear to be sensitive to mechanical damage. Therefore, pipet gently when dissociating the EpiLCs and transferring the cell suspension into a Lipidure-Coat Plate.

Acknowledgments

We would like to thank Drs. Hiroshi Ohta and Kazuki Kurimoto for their contribution to establishing this culture system. This study was supported, in part, by a Grant-in-Aid from the Ministry of Education, Culture, Sports, Science, and Technology of Japan; by PRESTO; by JST-CREST/ERATO; and by the Takeda Science Foundation. We apologize to researchers whose work has not been cited due to space limitations.

References

1. Lawson KA, Dunn NR, Roelen BA, Zeinstra LM, Davis AM, Wright CV et al (1999) Bmp4 is required for the generation of primordial germ cells in the mouse embryo. Genes Dev 13:424–436
2. Yabuta Y, Kurimoto K, Ohinata Y, Seki Y, Saitou M (2006) Gene expression dynamics during germline specification in mice identified by quantitative single-cell gene expression profiling. Biol Reprod 75:705–716

3. Kurimoto K, Yabuta Y, Ohinata Y, Shigeta M, Yamanaka K, Saitou M (2008) Complex genome-wide transcription dynamics orchestrated by Blimp1 for the specification of the germ cell lineage in mice. Genes Dev 22: 1617–1635
4. Seki Y, Hayashi K, Itoh K, Mizugaki M, Saitou M, Matsui Y (2005) Extensive and orderly reprogramming of genome-wide chromatin modifications associated with specification and early development of germ cells in mice. Dev Biol 278:440–458
5. Hajkova P, Ancelin K, Waldmann T, Lacoste N, Lange UC, Cesari F et al (2008) Chromatin dynamics during epigenetic reprogramming in the mouse germ line. Nature 452:877–881
6. Saitou M, Kagiwada S, Kurimoto K (2012) Epigenetic reprogramming in mouse preimplantation development and primordial germ cells. Development 139:15–31
7. Evans MJ, Kaufman MH (1981) Establishment in culture of pluripotential cells from mouse embryos. Nature 292:154–156
8. Martin GR (1981) Isolation of a pluripotent cell line from early mouse embryos cultured in medium conditioned by teratocarcinoma stem cells. Proc Natl Acad Sci U S A 78: 7634–7638
9. Brons IG, Smithers LE, Trotter MW, Rugg-Gunn P, Sun B, de Sousa C, Lopes SM et al (2007) Derivation of pluripotent epiblast stem cells from mammalian embryos. Nature 448:191–195
10. Tesar PJ, Chenoweth JG, Brook FA, Davies TJ, Evans EP, Mack DL et al (2007) New cell lines from mouse epiblast share defining features with human embryonic stem cells. Nature 448:196–199
11. Hubner K, Fuhrmann G, Christenson LK, Kehler J, Reinbold R, De La Fuente R et al (2003) Derivation of oocytes from mouse embryonic stem cells. Science 300: 1251–1256
12. Toyooka Y, Tsunekawa N, Akasu R, Noce T (2003) Embryonic stem cells can form germ cells in vitro. Proc Natl Acad Sci U S A 100:11457–11462
13. Geijsen N, Horoschak M, Kim K, Gribnau J, Eggan K, Daley GQ (2004) Derivation of embryonic germ cells and male gametes from embryonic stem cells. Nature 427:148–154
14. Nayernia K, Nolte J, Michelmann HW, Lee JH, Rathsack K, Drusenheimer N et al (2006) In vitro-differentiated embryonic stem cells give rise to male gametes that can generate offspring mice. Dev Cell 11:125–132
15. Hayashi K, Ohta H, Kurimoto K, Aramaki S, Saitou M (2011) Reconstitution of the mouse germ cell specification pathway in culture by pluripotent stem cells. Cell 146:519–532
16. Nichols J, Ying QL (2006) Derivation and propagation of embryonic stem cells in serum- and feeder-free culture. Methods Mol Biol 329:91–98
17. Ohinata Y, Sano M, Shigeta M, Yamanaka K, Saitou M (2008) A comprehensive, non-invasive visualization of primordial germ cell development in mice by the Prdm1-mVenus and Dppa3-ECFP double transgenic reporter. Reproduction 136:503–514

Chapter 14

Isolation and Culture of Porcine Neural Progenitor Cells from Embryos and Pluripotent Stem Cells

Mikkel A. Rasmussen, Vanessa J. Hall, and Poul Hyttel

Abstract

The isolation and culture of neural progenitor cells (NPCs) from pluripotent stem cells has facilitated in vitro mechanistic studies of diseases related to the nervous system, as well as discovery of new medicine. In addition, NPCs are envisioned to play a crucial role in future cell replacement therapy. The pig has become recognized as an important large animal model and establishment of in vitro-derived porcine NPCs would allow for preclinical safety testing by transplantation in a porcine biomedical model. In this chapter, a detailed method for isolation and in vitro culture of porcine NPCs from porcine embryos or induced pluripotent stem cells is presented. The neural induction is performed in coculture and the isolation of rosette structures is carried out manually to ensure a homogenous population of NPCs. Using this method, multipotent NPCs can be obtained in approximately 1 month. The cells have the potential of long-term culture and the ability to differentiate into neural and glial cells.

Key words Neural progenitor cells, Embryos, Induced pluripotent stem cells, Neurons and glia, Transplantation

1 Introduction

Neural progenitor cells (NPCs) are first formed during the development of the neural tube. These cells remain in the ventricular zone and later in the sub-ventricular zone and the sub-granular zone of the hippocampus, giving rise to new neurons and glia throughout life [1, 2]. The derivation and in vitro culture of NPCs from the rodent ventricular zone [3] and from mouse embryonic stem cells [4] has been reported long ago. When cultured in the presence of the mitogens, basic fibroblast growth factor (bFGF) and epidermal growth factor (EGF), NPCs show the same multipotent stem cell nature as their in vivo counterparts, including the capacity for long-term culture and the ability to differentiate into neural and glial cells in vitro.

Ramiro Alberio (ed.), *Epiblast Stem Cells: Methods and Protocols*, Methods in Molecular Biology, vol. 1074, DOI 10.1007/978-1-62703-628-3_14,

Dr. Thomson and colleague's establishment of human embryonic stem cells (hESC) in 1998 [5] and Dr. Yamanaka and Dr. Takahashi's generation of human induced pluripotent stem cells (iPSC) in 2007 [6], have allowed for the later derivation of NPCs from these human pluripotent stem cells [7, 8]. Differentiation of human iPSC-derived NPCs from patients with critical neurodegenerative diseases, including Alzheimer's disease, Parkinson's disease, and Amyotrophic lateral sclerosis into cortical neurons [9], dopaminergic neurons [10], or motor neurons [11], respectively, is currently being exploited as in vitro models of these incurable diseases. Although such in vitro models are vital for understanding the molecular mechanisms governing the diseases, as well as for discovery of new medicine, the ultimate goal of performing transplantation of iPSC-derived therapeutic cells will not be permitted in humans unless the procedures have been extensively tested in experimental animals.

The pig has become recognized as an important large animal model, which helps to bridge the gap between rodent and human studies, and transplantation of porcine in vitro-derived NPCs would provide important details on their regenerative potential in vivo. However, due to a lack of genuine ESCs in this species [12], NPCs have instead been derived directly from the inner cell mass [13] or the epiblast [14] of porcine embryos, and recently also from porcine iPSC (15; own unpublished results). In this chapter, a detailed method for isolation and culture of porcine NPCs from the epiblast or iPSCs is described. The neural induction is performed in coculture with murine stromal cells [16] and the isolation of rosette structures is carried out manually by needles to ensure a homogenous population of NPCs. Using this method, multipotent NPCs can be obtained in approximately 1 month. These cells have the potential for long-term culture in the presence of FGF and EGF and can differentiate into neural and glial cells.

2 Materials

Consumables were purchased from Sigma-Aldrich, St. Louis, MO, USA unless otherwise stated.

2.1 Consumables

- Thermometer.
- Styrofoam box.
- Scissors.
- Pean.
- 50 ml syringes.
- Flushing needle.

- Equivet uterine flushing catheter 65 cm (Kruuse #340846).
- 90 μm sterile filter.
- 3.5 cm culture dish.
- 4-well culture dishes.
- 6-well culture dishes.
- Glass coverslips.
- 15 ml falcon tubes.
- Insulin needles.
- Collagenase IV.
- Matrigel.
- TrypLE select (Life Technologies #12604-013).
- Sterile DMSO.
- Cryotubes.
- 4 % Paraformaldehyde (PFA).
- RNeasy mini kit.
- RevertAid first strand cDNA synthesis kit (Thermo Scientific #K1621).

2.2 Cells/Embryos

- Porcine in vivo Day 8–9 expanded, hatched blastocysts (obtained from inseminated sows).
- Porcine induced pluripotent stem cells (iPSCs; obtained by lentiviral reprogramming of porcine fibroblasts).
- Mitomycin-treated CF1 mouse embryonic fibroblast cells (MEF).
- MS5 stromal cells (DSMZ #441).

2.3 Media

Media were purchased from Life Technologies, (Paisley, UK) and growth factors were purchased from Prospec (Rehovot, Israel), unless otherwise stated.

Embryo flushing solution:

Dulbecco's phosphate-buffered saline (DPBS) without calcium and magnesium.

1 % fetal bovine serum (FBS).

Embryo isolation medium/Fibroblast medium:

Dulbecco's modified eagle's medium (DMEM).

10 % FBS.

2 mM L-glutamine.

1 % Penicillin/Streptomycin (Pen/strep).

Embryonic stem cell (ESC) medium:

DMEM/F12.

20 % knockout serum replacement (KSR).

1 % nonessential amino acids (NEAA).

0.2 % beta-mercaptoethanol.

20 ng/ml human bFGF.

1 % Pen/Strep.

Neural induction medium:

DMEM.

15 % KSR.

2 mM L-glutamine.

10 μM beta-mercaptoethanol.

250 ng/ml human Noggin.

1 % Pen/Strep.

Neural expansion medium:

DMEM/F12.

2 mM L-glutamine.

1 × B27 supplement.

1 × N2 supplement.

20 ng/ml human bFGF.

20 ng/ml human EGF.

1 % Pen/Strep.

Neural differentiation medium (first 2 weeks):

DMEM/F12.

1 × N2 supplement.

1 μM all-trans-retinoic acid (RA).

200 ng/ml human recombinant human sonic hedgehog (SHH).

20 ng/ml human recombinant brain-derived neurotrophic factor (BDNF).

0.2 mM ascorbic acid (AA).

1 % Pen/Strep.

Neural maturation medium (final week):

DMEM/F12.

1 × N2 supplement.

20 ng/ml human BDNF.

20 ng/ml human recombinant glial cell line-derived neurotrophic factor (GDNF).

0.2 mM AA.

1 % Pen/Strep.

2.4 Antibodies

Pluripotent stem cells:

- Oct3/4 Goat IgG 1:750 (Santa Cruz, SC-8628).

Neural progenitor cells:

- Nestin Rabbit IgG 1:4,000 (Millipore, AB5922).
- Vimentin Mouse IgG 1:500 (Zymed, 18-0052).

Neurons and glia:

- Beta-Tubulin III/TUJI Mouse IgG 1:4,000 (Sigma-Aldrich, T8660).
- GFAP Rabbit IgG 1:1,000 (Millipore, PAB5804).

2.5 Primers

List of primers used to analyze NPCs, neurons, and glia (Table 1).

3 Methods

An overview of the entire neural differentiation process including isolation of epiblasts, culture of pluripotent stem cells, derivation of NPCs and differentiation into mature neurons is depicted in Fig. 1. In general, culture of MEF feeder cells and MS5 stromal cells is carried out at 37 °C in an atmosphere of 20 % O_2 and 5 % CO_2, whereas, porcine epiblast outgrowth cultures, iPSCs, and NPCs are cultured at 38 °C in 5 % O_2 and 5 % CO_2. Medium is changed every second day and growth factors are added fresh on the day of the medium change. The medium stock is stable for up to 14 days at 4 °C.

3.1 Culture of Pluripotent Stem Cells

Day 1: Preparation of MEF feeder cells:

1. Thaw 1.2×10^6 Mitomycin C inactivated MEF feeder cells and plate 2×10^5 MEFs into each well of a gelatin-coated 6-well dish. Culture for 1 day in fibroblast culture medium.
2. The following day, check the confluence under the microscope, which should be close to 100 %. Rinse the MEFs once with DPBS and change the medium to ESC culture medium.

Day 0–5: Isolation and culture of epiblast cells:

1. Isolate the uterus from the abdominal cavity of sows immediately after slaughter at 8–9 days post insemination (Day 8–9).

Table 1
List of primers used to analyze neural progenitor cells, neurons and glia

Cell types	Primer name	Sequence (5′–3′)	Annealing temp.	Product size (bp)	Reference
NPCs	Nestin_Fw	GGCAGTGGTTCC AAGGCT	58	162	XM_001925549.1
	Nestin_Rv	GGCTGGCATAGG TGTGTCAA	59		
NPCs	Vimentin_Fw	GTGATGTCCGCC AGCAGT	57	218	DQ190948.1
	Vimentin_Rv	GCGTTCCAGAGA CTCGTT	53		
Neurons	TujI_Fw	GTGGTGCGGAA GGAGTGTG	59	218	FP102308.4
	TujI_Rv	TGGTGGATGGAC AGCGTGG	63		
Glia	GFAP_Fw	TTGACCTGCGACG TGGAGTC	61	225	[17]
	GFAP_Rv	AGGTGGCGATCT CGATGTCC	62		
Oligo-dendrocyte	MBP_Fw	GAGGCAGAGCTCC TGACTACAAA	61	101	[18]
	MBP_Rv	GTCCCGTCCT CCCAGCTT	59		
All	GAPDH_Fw	TCGGAGTGAA CGGATTTG	54	219	[19]
	GAPDH_Rv	CCTGGAAGATGG TGATGG	53		

2. Transfer the uterus to the cell culture laboratory in 38 °C warm water in a styrofoam box.
3. In the laboratory, remove the mesometrium from each uterine horn by use of a pair of scissors and isolate each horn by transection as close to the uterine body and cervix as possible.
4. Insert a uterine flushing catheter into the open cut cervical end of the uterine horn and a flushing needle into the oviduct and secure it by a pean (Fig. 2a).
5. Flush each uterine horn 2–3 times with a syringe containing 50 ml 38 °C embryo flushing solution.
6. Gently massage the uterus from the oviduct towards the catheter and collect the fluid in a glass bottle dispersed in a 38 °C warm water bath.

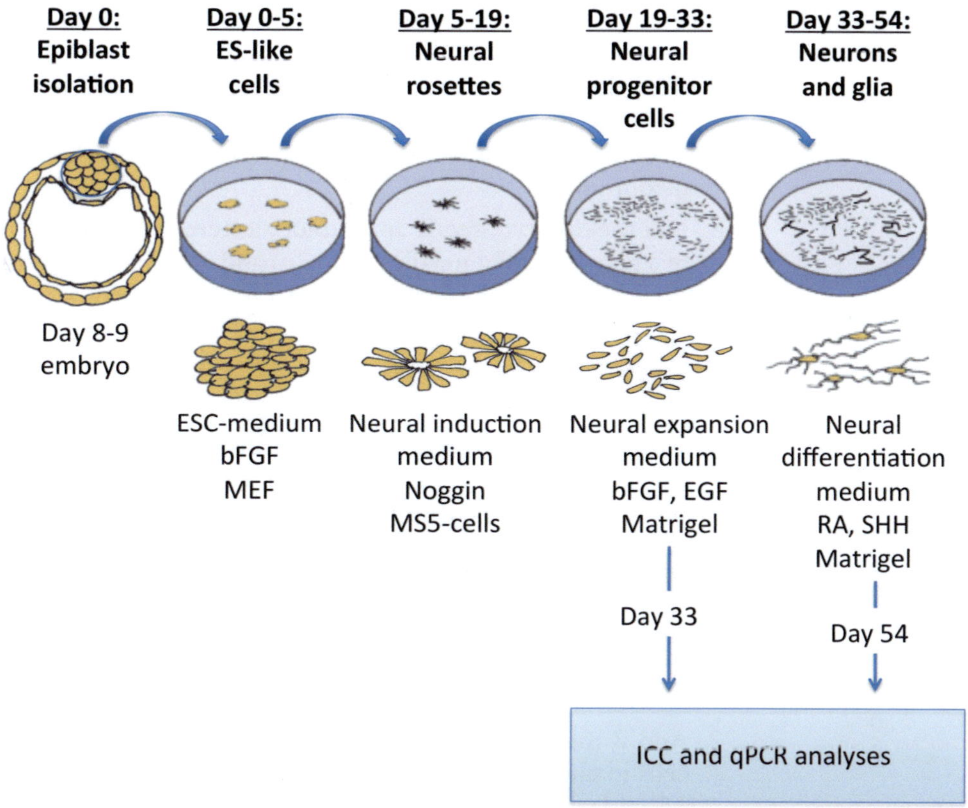

Fig. 1 Overview of the neural differentiation process. The sequential steps for derivation of neural progenitor cells (NPCs) from porcine epiblast cells. Culture conditions, including medium, growth factors and coating are indicated below

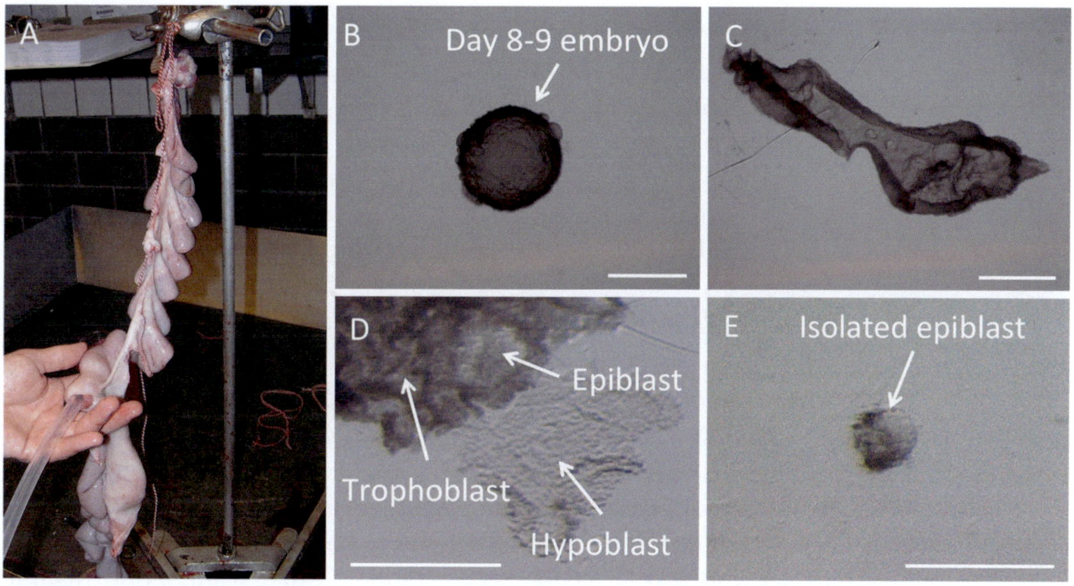

Fig. 2 Isolation of epiblasts. Isolation of epiblasts from Day 8 to 9 embryos. (**a**) The embryos are flushed from the uterine horn by an embryo flushing solution introduced into the oviduct and collected at open cut cervical end of the uterine horn via a flushing catheter. (**b**) Day 8–9 expanded hatched blastocyst. (**c**) Opening of Day 8–9 expanded hatched blastocyst by needles. (**d**) Isolation of the epiblast from the hypoblast and trophectoderm. (**e**) Isolated epiblast on MEF feeder cells. Scale bars represent 0.1 mm

7. Filter the solution through a 90 μm sterile filter and transfer the Day 8–9 expanded, hatched blastocysts (Fig. 2b) to a 3.5 cm culture dish containing 38 °C warm embryo isolation medium.
8. Pull the embryo gently apart under a stereomicroscope, by use of a pair of insulin needles (Fig. 2c) and identify the epiblast on the inside of the embryo (*see* **Note 1**).
9. Remove the hypoblast covering the epiblast (Fig. 2d) and gently dissect the epiblast away from the surrounding trophectoderm.
10. Transfer five isolated epiblasts (Fig. 2e) to each well of a 6-well dish containing MEF feeder cells and ESC medium by use of a 10 μl pipette and culture for 5 days.

Day 0–5: Culture of porcine induced pluripotent stem cells:

1. Rinse one 80 % confluent well of a 6-well dish with porcine iPSC once in DPBS and incubate for 5 min with 1 ml Collagenase IV.
2. Add 2 ml ESC medium to the well and gently scrape off the cells in clumps using a 5 ml pipette. Transfer the solution to a 15 ml falcon tube and spin at 1,200 × *g* for 5 min.
3. Replace the medium with 12 ml ESC medium and distribute it into each well of a 6-well dish (1:6 split) containing MEF feeder cells and culture for 5 days.

3.2 Establishment of NPCs

Day 4: Preparation of MS5 stromal cells:

1. Thaw 1.2×10^6 Mitomycin C inactivated MS5 stromal cells (*see* **Note 2**). Plate 2×10^5 MS5 cells into each well of a gelatin-coated 6-well dish and culture for 1 day in fibroblast medium.
2. On Day 5, check the confluence under the microscope, which should be close to 100 %. Rinse the MS5 stromal cells once with DPBS and change the medium to ESC medium.

Day 5–19: Neural induction:

1. On day 5, manually cut ESC-like areas of epiblast outgrowth colonies (Fig. 3a, c) or iPSCs (Fig. 3b) into 4–8 small pieces under the stereomicroscope by use of insulin needles.
2. Transfer the pieces from a single colony to a single well of a 6-well dish containing MS5 stromal cells by use of a 10 μl pipette and culture for 1 day in ESC medium.
3. On day 6, change the medium to neural induction medium (*see* **Notes 3** and **4**).

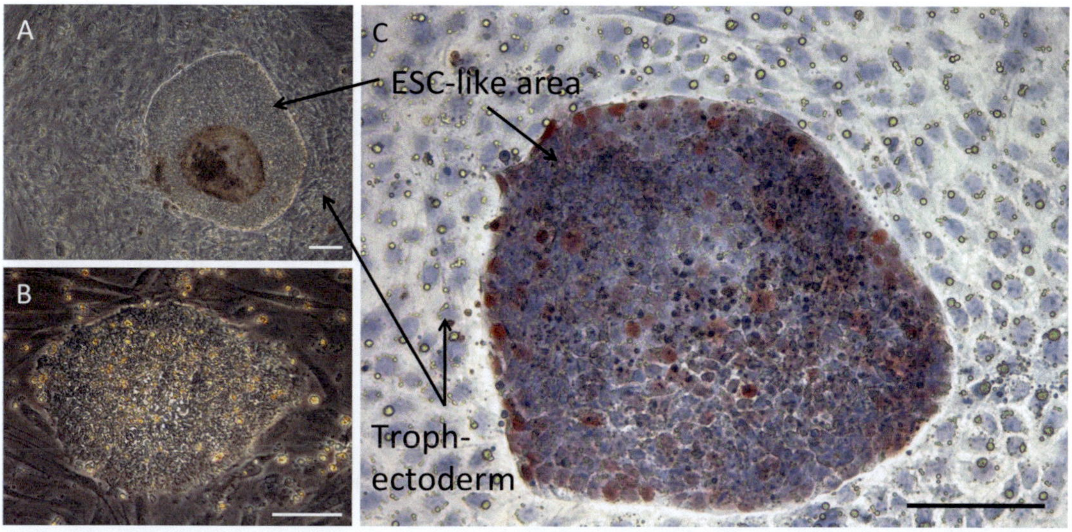

Fig. 3 Culture of pluripotent stem cells. In vitro culture of epiblasts or induced pluripotent stem cells (iPSC) on mouse embryonic feeder cells (MEF) cells. (**a**) Epiblast outgrowth colony cultured for 5 days on MEF cells. (**b**) Colony of induced pluripotent stem cells cultured for 5 days on MEF cells. (**c**). Chromogenic staining of an epiblast outgrowth colony cultured for 5 days on MEF cells with OCT3/4 (*red*) and hematoxylin (*blue*). *Arrows point* to ESC-like areas and trophectoderm cell areas. Scale bars represent 0.1 mm

Day 18: Preparation of Matrigel-coated dishes:

1. Thaw a 0.5 ml aliquot of Matrigel overnight at 4 °C.
2. Dilute with 15 ml cold DMEM/F12 and mix well.
3. Add 1 ml diluted Matrigel per well of a 6-well plate.
4. Incubate the Matrigel at room temperature for 1 h or overnight at 4 °C (*see* **Note 5**).
5. When ready to use, remove the excess liquid and wash once with DMEM/F12.

Day 19–33: Isolation of rosette structures and expansion of NPCs:

1. Manually cut rosette structures (Fig. 4a) into approximately four small pieces by use of insulin needles and isolate them from the surrounding cells (*see* **Notes 6** and **7**).
2. Transfer the isolated rosette structures from a single well of a 6-well dish (clonal line) to Matrigel-coated dishes and culture in neural expansion medium.
3. When a monolayer of NPCs is covering most of the dishes (Fig. 4b), rinse the cells once with DPBS and incubate for 5 min with 1 % TrypLE select at 37 °C to disaggregate the NPCs into single cells (*see* **Note 8**).
4. Add 2 ml neural expansion medium to each well, resuspend with a P1000 pipette and transfer to a 15 ml falcon tube and spin at 1,200 × g for 5 min.

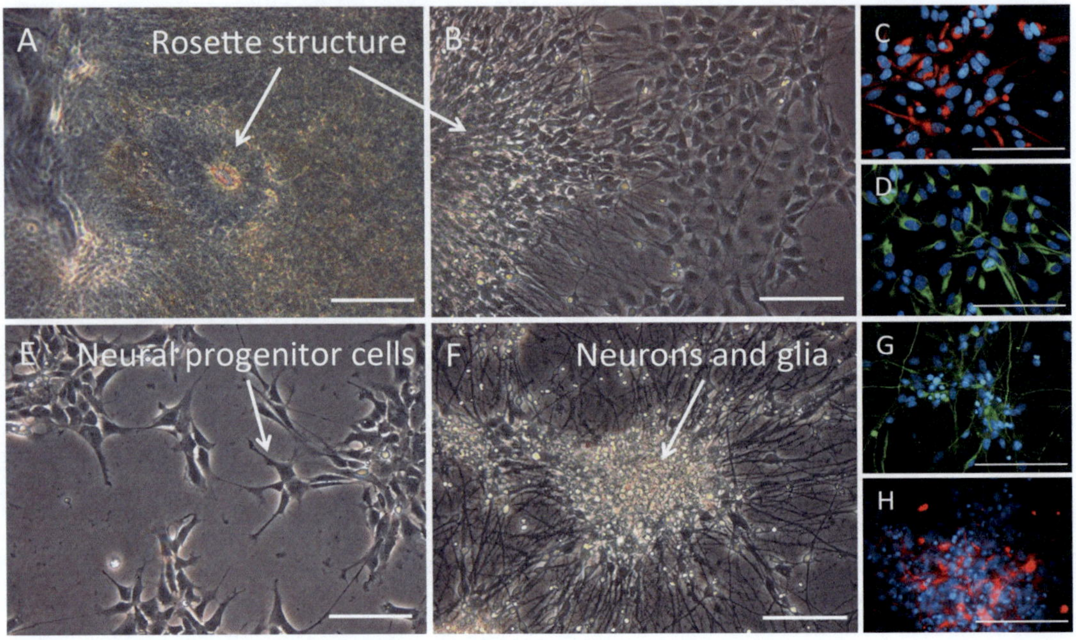

Fig. 4 Derivation of multipotent neural progenitor cells. Derivation of multipotent neural progenitor cells (NPCs) and differentiation to neurons and glia. (**a**) Neural rosette formed from culture of epiblast cells on MS5 feeder cells. *Arrow points* to a rosette structure. (**b**) Neural rosette isolated and cultured in Matrigel-coated dishes. (**c**) Immunocytochemical (ICC) staining of NPCs with NESTIN (*red*) and Hoechst (*blue*). (**d**) ICC staining of NPCs with VIMENTIN (*green*) and Hoechst (*blue*). (**e**) Neural progenitor cells cultured for 14 days on Matrigel-coated dishes. *Arrow points* to NPC cells. (**f**) Neuronal and glial-like morphology of NPCs cultured for 3 weeks in neural differentiation medium. (**g**) ICC staining of differentiated NPCs with TUJI (*green*) and Hoechst (*blue*). (**h**) ICC staining of differentiated NPCs with GFAP (*red*) and Hoechst (*blue*). Scale bars represent 0.1 mm

5. Replace the medium with 6 ml neural expansion medium and distribute the NPCs into three new Matrigel-coated wells of a 6-well dish (1:3 split) and culture for 5 days (termed passage 1; P1) (*see* **Note 9**).
6. Continue to split the NPCs (Fig. 4e) at a 1:3 to 1:5 ratio to new Matrigel-coated 6-well dishes with TrypLE select every 3–5 days and culture in neural expansion medium (*see* **Note 10**).
7. For immunocytochemical analysis, culture NPCs for 3–5 days in 4-well dishes with Matrigel-coated glass coverslips and fix in 4 % PFA for 20 min and store in DPBS at 4 °C.
8. For qPCR analysis, rinse NPCs from three wells of a 6-well dish once with DPBS, add 750 μl RLT lysis buffer to each well and resuspend with a P1000 pipette, transfer to RNA-free eppendorf tubes and store at −80 °C.

3.3 Freezing

1. Label cryotubes with NPC cell line number, passage number and current date.

2. Rinse one well of a 6-well dish once with DPBS and incubate for 5 min with 1 % TrypLE select at 37 °C to disaggregate the NPCs into single cells.
3. Add 2 ml neural expansion medium to each well, resuspend with a P1000 pipette and transfer to a 15 ml falcon tube and spin at 1,200 × *g* for 5 min.
4. Replace the medium with 1 ml prechilled neural expansion medium containing 10 % DMSO and freeze the NPCs overnight at −80 °C.
5. Transfer the vials into liquid nitrogen the day after for long-term storage.

3.4 Thawing

1. Thaw the NPCs in a 38 °C water bath until a small ice clump remains.
2. Wipe the tube with ethanol and transfer the cells into a 15 ml falcon tube containing 6 ml pre-warmed DMEM/F12 and spin at 1,200 × *g* for 5 min.
3. Replace the medium with 2 ml neural expansion medium and transfer to one Matrigel-coated well of a 6-well dish (1:1).

3.5 Evaluation of Neural Differentiation Potential

Day 33–54: Neural differentiation of NPCs:

1. Disaggregate the NPCs into single cells using TrypLE select and seed cells at a density of $1 \times 10^5/\text{cm}^2$ in Matrigel-coated 6-well dishes and 4-well dishes containing Matrigel-coated glass coverslips.
2. Culture the NPCs for 2 weeks in neural differentiation medium without any further cell passaging.
3. After 2 weeks, change the medium to neural maturation medium and culture for 1 additional week without passaging (Fig. 4f).
4. For immunocytochemical analysis, fix neurons cultured in 4-well dishes with Matrigel-coated glass coverslips in 4 % PFA for 20 min and store in DPBS at 4 °C.
5. For qPCR analysis, rinse neurons from three wells of a 6-well dish once with DPBS, add 750 μl RLT lysis buffer to each well and resuspend with a P1000 pipette, transfer to RNA-free eppendorf tubes and store at −80 °C.

3.6 Analysis

Immunocytochemistry:

1. Remove DPBS and permeabilize NPCs and neurons cultured on Matrigel-coated glass coverslips for 30 min in 0.1 % Triton X-100 in DPBS.
2. Block for 1 h in 5 % Donkey serum in PBS.

3. Incubate with primary antibodies diluted in 0.25 % BSA, 0.1 % Triton X-100 in DPBS overnight at 4 °C.
4. As negative controls, perform additional incubations with a mouse IgG or IgM antibody in the same concentration as the primary antibodies (primary antibody control). As an additional negative control, perform incubation without the primary antibody (secondary antibody control).
5. The following day, wash cells three times in DPBS.
6. Incubate for 1 h with fluorescent-conjugated secondary antibodies (Alexa Fluor; 1:400) diluted in 0.25 % BSA, 0.1 % Triton X-100 in DPBS.
7. Wash three times in DPBS and incubate in 0.1 μl/ml Bisbenzimide Hoechst
8. Add a drop of fluorescence mounting medium onto a glass slide and mount glass coverslips upside-down directly onto the glass slides.
9. Visualize the antibody stainings of NPCs (Fig. 4c, d) and neurons (Fig. 4g, h) by use of a fluorescent microscope.

Comparative real-time PCR:

1. Purify total RNA from NPCs and neurons and a reference tissue (e.g. porcine brain from Day 42 fetus) by use of the RNeasy mini kit according to the manufacturer's instructions and measure the RNA content and purity on a spectrophotometer.
2. Perform reverse transcription using a RevertAid first strand cDNA synthesis kit according to the manufacturer's instructions.
3. Setup a qPCR plate with technical and biological triplicates of cDNA from cells and reference tissue and porcine-specific primers as target genes and glyceraldehyde 3-phosphate dehydrogenase (GAPDH) as a reference gene [19].
4. Perform comparative real-time PCR on a qPCR machine with the following PCR conditions: 45 cycles of denaturation at 95 °C for 10 s, annealing at 58 °C for 10 s and elongation at 72 °C for 20 s.
5. Use the $2^{-\Delta\Delta CT}$ method to calculate the difference in expression between the samples.

4 Notes

1. The epiblast is glass-clear and displays a characteristic almost lens-like reflection of the light, whereas the trophectoderm appears darker. The hypoblast on the inside of the embryo may, due to its fixation to the epiblast, guide to the position of the epiblast.

2. The MS5 stromal cells should be thawed and subcultured to passage 5. The cells are Mitomycin C inactivated prior to freezing in aliquots of 1.2×10^6 (sufficient for coating of one 6-well dish).
3. In order to establish clonal NPC lines from single embryos it is vital to culture the epiblasts separately throughout the procedure, especially during neural induction, when individual epiblast outgrowth colonies are cut into smaller pieces.
4. Rosette structures should typically begin to appear after 7–14 days in coculture (Days 12–19).
5. Plates may either be used immediately or stored at 4 °C for up to 2 weeks.
6. Rosette structures are characterized as round structures composed of radially organized columnar epithelial cells, which mimic the neuroepithelial structures of anterior CNS fate.
7. Isolation of rosette structures should be performed with great care, as transfer of additional cells surrounding the rosette structure may contaminate the cultures.
8. NPCs are characterized as small bipolar cells with short processes, which proliferate extensively in culture.
9. It is important to culture the NPCs in high confluence, especially in the initial cultures, which promotes their proliferation and survival.
10. The NPCs can proliferate undifferentiated for more than 30 passages, although their proliferation may decrease over time. Long-term culture can change the differentiation potential of the NPCs so that more glial cells are produced with higher passage number.

Acknowledgement

The Danish National Advanced Technology Foundation (HTF) 047-2011-1, The Danish Council for Independent Research, Natural Sciences (FNU), 11-106627, and The Danish Council for Strategic Research (DSF)—The Program Commission on Individuals, Disease and Society.

References

1. Temple S (2001) The development of neural stem cells. Nature 414:112–117
2. Conti L, Pollard SM, Gorba T, Reitano E, Toselli M, Biella G, Sun Y, Sanzone S, Ying QL, Cattaneo E, Smith A (2005) Niche-independent symmetrical self-renewal of a mammalian tissue stem cell. PLoS Biol 3:283
3. Temple S (1989) Division and differentiation of isolated CNS blast cells in microculture. Nature 340:471–473
4. Okabe S, Forsberg-Nilsson K, Spiro AC, Segal M, McKay RD (1996) Development of neuronal precursor cells and functional postmitotic neurons from embryonic stem cells in vitro. Mech Dev 59:89–102

5. Thomson JA, Itskovitz-Eldor J, Shapiro SS, Waknitz MA, Swiergiel JJ, Marshall VS, Jones JM (1998) Embryonic stem cell lines derived from human blastocysts. Science 282: 1145–1147
6. Takahashi K, Tanabe K, Ohnuki M, Narita M, Ichisaka T, Tomoda K, Yamanaka S (2007) Induction of pluripotent stem cells from adult human fibroblasts by defined factors. Cell 131:861–872
7. Reubinoff BE, Itsykson P, Turetsky T, Pera MF, Reinhartz E, Itzik A, Ben-Hur T (2001) Neural progenitors from human embryonic stem cells. Nat Biotechnol 19:1134–1140
8. Chambers SM, Fasano CA, Papapetrou EP, Tomishima M, Sadelain M, Studer L (2009) Highly efficient neural conversion of human ES and iPS cells by dual inhibition of SMAD signaling. Nat Biotechnol 27:275–280
9. Shi Y, Kirwan P, Smith J, MacLean G, Orkin SH, Livesey FJ (2012) A human stem cell model of early Alzheimer's disease pathology in Down syndrome. Sci Transl Med 4:124ra29
10. Sánchez-Danés A, Richaud-Patin Y, Carballo-Carbajal I, Jiménez-Delgado S, Caig C, Mora S, Di Guglielmo C, Ezquerra M, Patel B, Giralt A, Canals JM, Memo M, Alberch J, López-Barneo J, Vila M, Cuervo AM, Tolosa E, Consiglio A, Raya A (2012) Disease-specific phenotypes in dopamine neurons from human iPS-based models of genetic and sporadic Parkinson's disease. EMBO Mol Med 4: 380–395
11. Dimos JT, Rodolfa KT, Niakan KK, Weisenthal LM, Mitsumoto H, Chung W, Croft GF, Saphier G, Leibel R, Goland R, Wichterle H, Henderson CE, Eggan K (2008) Induced pluripotent stem cells generated from patients with ALS can be differentiated into motor neurons. Science 321:1218–1221
12. Wolf XA, Rasmussen MA, Schauser K, Jensen AT, Schmidt M, Hyttel P (2011) OCT4 expression in outgrowth colonies derived from porcine inner cell masses and epiblasts. Reprod Domest Anim 46:385–392
13. Puy L, de Sousa C, Lopes SM, Haagsman HP, Roelen BA (2010) Differentiation of porcine inner cell mass cells into proliferating neural cells. Stem Cells Dev 19:61–70
14. Rasmussen MA, Hall VJ, Carter TF, Hyttel P (2011) Directed differentiation of porcine epiblast-derived neural progenitor cells into neurons and glia. Stem Cell Res 7:124–136
15. Yang JY, Mumaw JL, Liu Y, Stice SL, West FD (2013) SSEA4 positive pig induced pluripotent stem cells are primed for differentiation into neural cells. Cell Transplant 22(6):945–959
16. Perrier AL, Tabar V, Barberi T, Rubio ME, Bruses J, Topf N, Harrison NL, Studer L (2004) Derivation of midbrain dopamine neurons from human embryonic stem cells. Proc Natl Acad Sci U S A 101:12543–12548
17 Dyce PW, Zhu H, Craig J, Li J (2004) Stem cells with multilineage potential derived from porcine skin. Biochem Biophys Res Commun. Apr 9;316(3):651–8. PubMed PMID: 15033449
18 Zeng L, Rahrmann E, Hu Q, Lund T, Sandquist L, Felten M, O'Brien TD, Zhang J, Verfaillie C. Multipotent adult progenitor cells from swine bone marrow. Stem Cells. 2006 Nov;24(11):2355–66. Epub 2006 Aug 24. Erratum in: Stem Cells. 2007 Sep;25(9):2387. PubMed PMID: 16931778
19 Kuijk EW, du Puy L, van Tol HT, Haagsman HP, Colenbrander B, Roelen BA. Validation of reference genes for quantitative RT-PCR studies in porcine oocytesand preimplantation embryos. BMC Dev Biol. 2007 May 31;7:58. PubMed PMID: 17540017; PubMed Central PMCID: PMC1896162

Chapter 15

Derivation of Neural Precursors from Bovine Preimplantation Embryos

Giovanna Lazzari, Silvia Colleoni, and Cesare Galli

Abstract

Induction of neural differentiation from embryonic pluripotent stem cells (ES/EpiSc), both of mouse and primates, has been extensively published by several research teams. However, direct derivation of organized neuroectoderm in vitro from blastocyst stage embryos of rodents or primates has not been reported so far. Here we describe a method of direct neural differentiation from the inner cell mass cells of preimplantation bovine embryos, without the intermediate step of ES/EpiSc cells derivation (Lazzari et al., Stem cells 24:2514–2521, 2006). Proliferating neural precursors cells lines, and both central and peripheral nervous system derivatives, can be obtained providing a unique in vitro model of early neurulation events in mammals.

Key words Preimplantation embryo, Bovine, Neural differentiation, Neural rosettes, Neural precursors

1 Introduction

In vertebrate embryos the process of neurulation starts with the folding of the dorsal ectoderm that gives rise to the inner neural tube and the outer epidermis while, at the same time, the neural crest delaminates in between the two [1, 2]. This early gastrulation process has been partially reproduced in vitro by inducing the neural differentiation of mouse and human ES cells [3], using protocols based on serum-free media or coculture with stromal feeders producing inductive signals [4–8]. Contrary to these differentiation protocols, that are used on established ES or EpiSc cell lines, the method described in this chapter is applied directly to the inner cell mass cells isolated from bovine preimplantation embryos, providing a unique model for the study of mammalian neurulation in vitro, with specific emphasis on the neural crest lineage.

Under neural differentiation conditions, the bovine inner cell mass cells undergo uniform and reproducible morphological

Ramiro Alberio (ed.), *Epiblast Stem Cells: Methods and Protocols*, Methods in Molecular Biology, vol. 1074, DOI 10.1007/978-1-62703-628-3_15, © Springer Science+Business Media New York 2013

and functional changes that give rise to colonies of neural rosettes resembling the developing neural tube [9]. These typical structures have also been described following differentiation of primate ES cells [5, 10], while differentiating rodent ES and EpiSc express neural markers, but fail to organize in well formed, radially shaped, rosettes. The radial organization of the differentiated cells is also functionally similar to the neural tube as demonstrated by the typical feature of interkinetic nuclear migration, whereby rosette cells nuclei migrate at center or at the periphery according to their cell cycle stage. Another feature is the presence of peripheral P75 positive cells, the precursors of the delaminating neural crest lineage. Finally, rosette cells can respond to inductive signals similar to the cells of the developing neural tube, changing their dorsal identity to ventral and posterior fates according to the exposure to morphogens such as SHH and retinoic acid.

Following trypsinization and replating in feeder-free conditions, in the presence of bFGF and EGF, rosette cells give rise to neural precursor cell lines capable of long-term proliferation (>120 doublings). The predominant default differentiation, in the absence of specific inductive signals, is anterior and dorsal, as demonstrated by the expression of the forebrain (Otx1, Pax6, Emx2), midbrain (Otx1, Dmbx1), and hindbrain (Hoxb1) markers and the neural crest marker p75, while spinal cord-specific genes (Hoxb2, Hoxb4, and Hoxb9) are not expressed. Following growth factors withdrawal the proliferating precursors undergo terminal differentiation into a variety of cell types that comprise all the derivatives of the neural crest lineage (autonomic and sensory peripheral neurons, glial cells, melanocytes, smooth muscle cells, and condrocytes). In the presence of inductive signals such as Shh, the rosette cells acquire a central nervous system (CNS) identity and the derived CNS precursors express floor plate and ventral markers such as FoxA2, Shh, Isl1, Nkx2.2, and Nkx6.1, and differentiate into mature astrocytes and neurons of CNS ventral identity [9].

This chapter describes in detail the method summarized above for the direct induction of neural differentiation from bovine blastocysts and for the derivation and differentiation of neural precursor cell lines.

The first section deals with common methods for the production of bovine embryos in vitro, the second section deals with the culture of isolated inner cell mass cells, the induction of neural differentiation, and the formation of neural rosettes. The third section describes the derivation of neural precursors cell lines and their terminal differentiation.

2 Materials

2.1 Production of Bovine Embryos

Media formulation for production of bovine embryos.

All media are adjusted to mOsm 280–290 and pH 7.35–7.45 (after equilibration in 5 % CO_2 incubator, unless differently indicated).

- TCM199-HEPES: TCM199 (Sigma), 4 g/l BSA, 0.05 g/l heparin, 1.68 g/l sodium bicarbonate, 0.35 g/l penicillin, 0.35 g/l streptomycin, 2.6 g/l NaOH.
- TCM199-IVM: TCM199 (Sigma), 10 % (v/v) fetal calf serum, 2.2 g/l sodium bicarbonate, 10 ml/l Glutamax (Life Technologies), 1 ml/l ITS Media Supplement, 1 mM sodium pyruvate, 0.5 mM L-cystein, 10 mM glycine, 75 mg/l Kanamycin, 0.05 IU/ml Follicle Stimulating Hormone, and 0.05 IU/ml Luteinizing Hormone (Menopur, Ferring Pharmaceuticals, Inc.), 50 ng/ml long-EGF human (Sigma), 100 ng/ml long-IGF-I human (Sigma), and 10 ng/ml bFGF.
- TALP calcium free 1×: 310 mg/l $MgCl_2 \cdot 6H_2O$; 6 g/l NaCl, 230 mg/l KCl, 420 mg/l $NaHCO_3$, 40 mg/l Na_2HPO_4, 110 mg/l sodium pyruvate, 4,760 mg/l HEPES, 75 mg/l kanamycin, 3.7 ml/l sodium lactate 60 % (Sigma), 400 mg/l NaOH.
- TALP calcium free 10×: 310 mg/100 ml $MgCl_2 \cdot 6H_2O$; 6 g/100 ml NaCl, 230 mg/100 ml KCl, 420 mg/100 ml $NaHCO_3$, 40 mg/100 ml Na_2HPO_4, 110 mg/100 ml sodium pyruvate, 4,760 mg/100 ml HEPES, 75 mg/100 ml kanamycin, 3.7 ml/100 ml sodium lactate 60 %, 300 mg/100 ml NaOH.
- TALP-IVF: 390 mg/l $CaCl_2 \cdot 2H_2O$, 100 mg/l $MgCl_2 \cdot 6H_2O$, 6,200 mg/l NaCl, 230 mg/l KCl, 2,200 mg/l $NaHCO_3$, 47 mg/l Na_2HPO_4, 110 mg/l sodium pyruvate, 75 mg/l kanamycin, 1.86 ml/l sodium lactate 60 %, 6 g/l BSA FAF, 1 mg/l heparin, 20 μM D-penicillamine, 100 μM hypotaurine, 1 μM epinephrine, 4 ml/l NaOH 1 N.
- TALP-WASH: 390 mg/ml $CaCl_2 \cdot 2H_2O$, 100 mg/l $MgCl_2 \cdot 6H_2O$; 7,300 mg/l NaCl, 235 mg/l KCl, 168 mg/l $NaHCO_3$, 47 mg/l Na_2HPO_4, 110 mg/l sodium pyruvate, 2,380 mg/l HEPES, 75 mg/l kanamycin, 1.9 ml/l sodium lactate 60 %, 6 g/l BSA FAF.75 ml/l NaOH 1 N.
- TCM199-IVC: TCM199, 16 g/l BSA FAF, 10 mM Glycine, 36 mg/l sodium pyruvate, 1 ml NaOH 1 N.
- 90 % Redigrad solution: 9 ml Redigrad (Amersham) added to 1 ml of medium TALP calcium free 10×.
- 45 % Redigrad solution: 1 ml of 90 % Redigrad solution added to 1 ml of medium TALP calcium free 1×.

2.2 Media Formulation for Culture of ICM Cells Isolated from Bovine Embryos (DMEM/F12-Plating)

- Medium DMEM/F12, 15 % Knockout Serum Replacement (Invitrogen), 2 mM glutamine, 100 μM β-mercaptoethanol, with or without 2 ng/ml bFGF.

2.3 Media Formulation for Culture of Bovine Neural Precursors (DMEM/F12-Neural)

- Medium DMEM/F12, 0.6 % glucose, 3 mM sodium bicarbonate, 2 mM glutamine, 5 mM HEPES, 25 μg/ml insulin, 60 μM putrescine, 20 nM progesterone, 100 μg/ml transferrin, 30 nM sodium selenite, 2 μg/ml heparin, 10 ng/ml bFGF, 20 ng/ml EGF. mOsm 330–340, pH 7.2–7.3.
- Matrigel solution: Matrigel diluted 1:100 with DMEM-F12.
- Trypsin inhibitor solution: 1 g/l Trypsin inhibitor in PBS.
- Collagenase solution: 1 mg/ml collagenase IV in TCM199-HEPES.
- STO fibroblasts from American Type Culture Collection (CRL-1503).
- Glass petri dish.

2.4 Immunosurgery

- TCM199-PVA: TCM199, 1.68 g/l sodium bicarbonate, 0.35 g/l penicillin, 0.35 g/l streptomycin, NaOH 2.6 g/l, 0.1 % polyvinyl alcohol.
- TCM199-FCS: TCM199, 1.68 g/l sodium bicarbonate, 0.35 g/l penicillin, 0.35 g/l streptomycin, NaOH 2.6 g/l, 10 % fetal calf serum.
- TCM199-antiserum: TCM199, 1.68 g/l sodium bicarbonate, 0.35 g/l penicillin, 0.35 g/l streptomycin, NaOH 2.6 g/l, 20 % bovine antiserum.
- TCM199-GPC: TCM199, 1.68 g/l sodium bicarbonate, 0.35 g/l penicillin, 0.35 g/l streptomycin, NaOH 2.6 g/l, 10 % Guinea pig complement.
- Pronase solution: 0.1 % pronase in PBS.

3 Methods

3.1 Production of Bovine Embryos

3.1.1 Recovery of Bovine Ovaries and Collection of Oocytes

1. Bovine ovaries are recovered at the abattoir from pubertal females subjected to routine veterinary inspection and in accordance to the specific health requirements stated in Council Directive 89/556/ECC and subsequent modifications.
2. The ovaries are transported to the laboratory in a thermostatic container at 25 °C within 4 h from the death of the animals.
3. In the laboratory the ovaries are rinsed twice with sterile PBS.

4. Bovine oocytes are recovered from the ovarian follicles with the aid of a needle (Surflo, winged infusion set—19G, Terumo) connected to an aspiration pump (Hirshmann Laborgerate). The recovered oocytes are washed in medium TCM199-HEPES. Only oocytes surrounded with a homogenous compact cumulus are selected for maturation.

3.1.2 In Vitro Maturation and In Vitro Fertilization

1. The selected oocytes are transferred to maturation medium composed of medium TCM 199-IVM. Maturation is carried out at 38.5 °C in 5 % CO_2 in humidified air for 18–22 h.
2. Two hours before the completion of maturation frozen bovine semen is thawed and layered on top of a discontinuous Redigrad gradient composed of two fractions: 90 % (2 ml, at the bottom) and 45 % (2 ml) Redigrad in a centrifuge tube.
3. The semen is centrifuged at 740 × *g* for 40 min.
4. The motile fraction is recovered at the bottom of the tube (approx. 150 microl.), transferred to a second centrifuge tube, diluted with TALP calcium free (4 ml) and centrifuged at 370 × *g* for 10 min.
5. The sperm pellet is counted and diluted at a final concentration of 0.5–1.5 million sperm/ml, depending on the quality of the semen used for in vitro fertilization. The medium used for sperm dilution is TALP-IVF.
6. Matured oocytes are washed in medium TCM199 and added to 300 μl of sperm suspension dispensed in 4-well plates.

3.1.3 Embryo Culture

1. After co-incubation for 18–20 h with the sperm the presumptive zygotes are denuded of cumulus cells by vortexing.
2. The denuded presumptive zygotes are transferred in medium SOFaa [11] in a final volume of 400 μl in 4-well plates.
3. Half of the medium is changed on day 4 (fertilization = day 0) with the same medium.
4. Half of the medium is changed on day 6 with medium TCM199-IVC.
5. Bovine blastocysts form on day 7–8.

3.2 Derivation of Neural Rosettes from Bovine Embryos

1. Seven to eight days after in vitro fertilization the expanded or peri-hatching blastocysts are recovered from the culture well. Only embryos with a prominent and well-defined inner cell mass (ICM) are suitable for the derivation of neural rosettes (Fig. 1a) (*see* **Note 1**).
2. The zona pellucida is removed with pronase and the embryos are either dissected manually using 25 G needles or treated by immunosurgery to isolate the ICM.

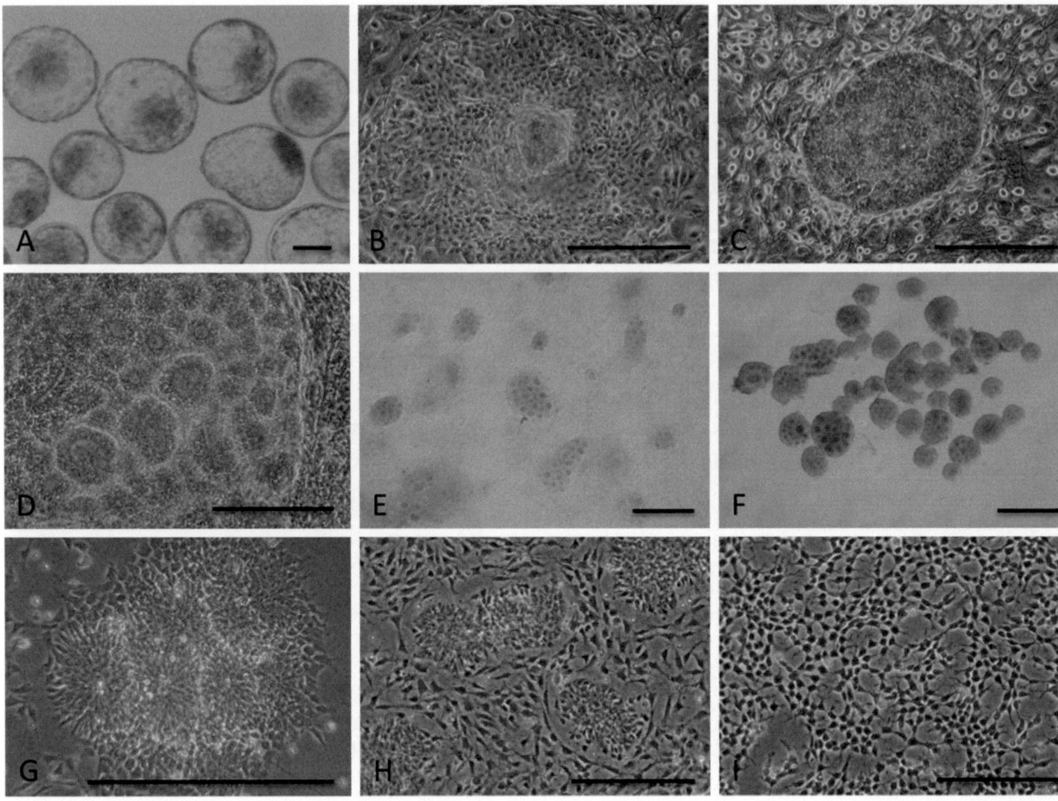

Fig. 1 (**a**) Bovine embryos at the expanded peri-hatching stage, 7 days after in vitro fertilization; (**b**) outgrowth 6 days post plating of the inner cell mass: clearly visible epiblast covered and surrounded with a layer of primitive endoderm; (**c**) primary colony formed after removal of the primitive endoderm and replating; (**d**) colony of neural rosettes; (**e**) idem, view at the stereomicroscope; (**f**) colonies of neural rosettes detached from the feeders by enzymatic treatment; (**g**) neural rosettes plated on matrigel; (**h–i**) neural precursor cell line at passage 3 and 10, respectively. Bars 100 μm

3. The isolated ICMs are plated on mitomycin-inactivated feeders STO fibroblasts (*see* **Note 2**) in medium DMEM/F12-plating with or without supplementation of bFGF (*see* **Note 3**).
4. One week later the outgrowths (Fig. 1b) (*see* **Note 4**) are disaggregated with 0.05 % trypsin–EDTA and replated (*see* **Note 5**).
5. During the next 2–3 days the colonies become more visible with very neat margins. The cells forming the colonies begin to organize in small radial structures that progressively acquire the typical morphology of neural rosettes (Fig. 1d, e).
6. Seven to ten days after passaging the rosettes-containing colonies they are separated from STO feeders by brief exposure (10–20 min) to collagenase (Fig. 1f). They are cut in small fragments with insulin needles, over a glass petri dish and plated on matrigel (1:100 dilution) coated dishes in medium DMEM/F12-neural.

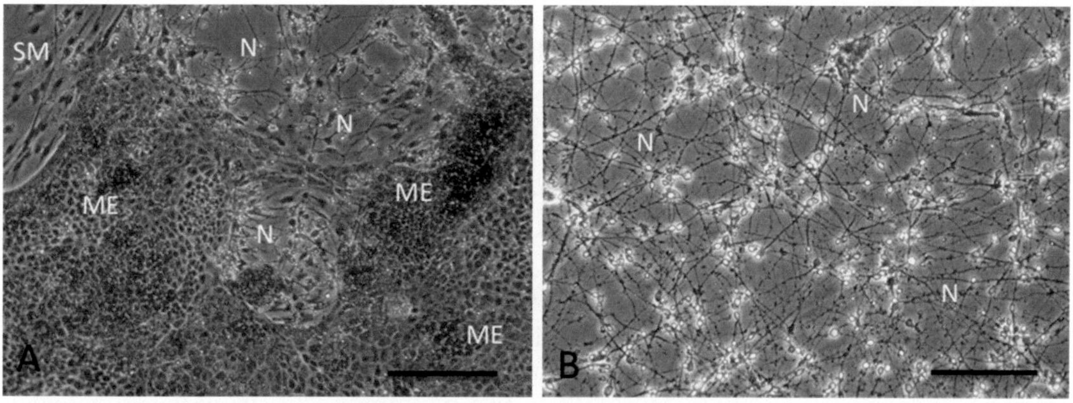

Fig. 2 (**a**) Differentiated neural crest precursors 20 days following growth factors withdrawal (ME: melanocytes, N: neurons; SM: smooth muscle cells). (**b**) Differentiated neural precursors committed to neuronal fate only, 20 days following growth factors withdrawal (N: neurons). Bars 50 μm

3.2.1 Immunosurgery

1. The zona pellucida is removed from the embryos if it is still present by exposure to pronase.
2. The embryos are rinsed in TCM199-PVA and transferred to TCM 199-antiserum at 37–39 °C and incubated for 20 min.
3. The embryos are transferred in TCM199-FCS and incubated for 10 min.
4. The embryos are transferred in TCM199-GPC and incubated for 20 min.
5. The embryos are transferred to TCM199-FCS and with a fine bore pipette the debris of lysed trophoblast cells are removed, leaving a clean cluster of inner cell mass cells.

3.3 Derivation and Characterization of Neural Precursors Cell Lines

1. Following plating onto matrigel the fragments of the colonies flatten, and the typical rosette morphology is clearly distinguishable (Fig. 1g).
2. Following passages by trypsinization (*see* **Note 6**) the rosette structure is progressively lost and the cultures become homogeneous (Fig. 1h, i).
3. The derived cell lines can be passaged for several times up to over 120 population doublings (*see* **Note 7**).
4. The neural precursors cell lines at the undifferentiated stage, express markers of the neural crest: Slug, Sox10, and FoxD3.
5. To induce differentiation, growth factors are removed from the culture medium. Under these conditions the cells differentiate extensively into all the variety of cell types that comprise the neural crest lineage. These include autonomic and sensory peripheral neurons, glial cells, melanocytes, smooth muscle cells, and condrocytes (Fig. 2a) [9].

6. If differentiation is induced at advanced passage number (over 10–15), although the morphology of the undifferentiated cultures remains stable, the ability for differentiation becomes restricted to condrocytes and smooth muscle cells, while very few neurons can be identified in the cultures.
7. In some instances the cell lines have a mixed identity of neural crest and CNS. They become obvious at the time of induction of differentiation when they give rise to an almost exclusive neuronal differentiation (Fig. 2b).
8. The neural prercursors cell lines and their differentiated derivative can be extensively characterized with a panel of markers, both by RT-PCR and by immunocitochemistry (supplementary table 1 in [5]).

4 Notes

1. Embryo selection is crucial for achieving consistent results. Embryos should be scored not simply for having reached the blastocyst stage, but also for the presence of a well-defined ICM, the absence of degenerating blastomeres and general uniform and relatively clear morphology. Dark embryos, even having a large ICM, are indicative of lipid accumulation and metabolic stress. These embryos do not provide high-quality ICM clusters after immunosurgery.
2. STO feeders are very sensitive to overgrowth and if the cultures are allowed to become overcrowded (more than two millions cells per 6 cm dish) very quickly the STO cells lose contact inhibition and tend to overgrow even after mitomycin treatment. In that case the culture must be discarded and a new stock must be thawed.
3. The supplementation of bFGF accelerates and increases the size of the outgrowth, but is not absolutely required for the growth of the plated inner cell mass.
4. The outgrowths that derive from large ICMs without any contamination of trophoblast cells can acquire a typical "fried egg" appearance that is given by the development of a layer of primitive endoderm that covering the cluster of pure epiblast cells (Fig. 1b). In this case it is helpful to remove the layer of endoderm by lifting it using a fire-pulled pasteur pipette. This procedure is facilitated by a brief exposure (10–20 min) of the outgrowth to collagenase. In this case the cluster of epiblast cells is replated without disaggregation that is postponed to the following passage, 4–6 days later.
5. The disaggregation of the outgrowth is best performed in microdrops of 40 μl. After 5–8 min exposure to the enzyme,

10 μl of trypsin inhibitor must be added to the drop. Soon afterwards the outgrowth must be transferred to another drop (of the same volume) of DMEM/F12-plating medium using a micro-pipette calibrated at 5 μl. The outgrowth is then disaggregated with the aid of a fire-pulled pasteur pipette and replated on fresh feeders. For larger outgrowths the disaggregation can be achieved by repeated pipetting using a micropipette calibrated at 20 μl. In this case, following disaggregation, the whole volume of the microdrop can be transferred in a new culture well with inactivated feeders and fresh medium.

6. Serum-free cultures require more care during trypsination avoiding excess of exposure (over 5 min) and pipetting. The use of a trypsin inhibitor is imperative to inactivate trypsin.
7. The recommended passaging regime is every 3/4 days and the plating density is 20,000 cells/cm^2. Doubling time varies between 22 and 30 h, depending on the cell line. Leaving the cells at high density for additional days promotes spontaneous differentiation.

References

1. Knecht AK, Bronner-Fraser M (2002) Induction of the neural crest: a multigene process. Nat Rev Genet 3:453–461
2. Copp AJ, Greene ND, Murdoch JN (2003) The genetic basis of mammalian neurulation. Nat Rev Genet 4:784–793
3. Gaspard N, Vanderhaeghen P (2010) Mechanisms of neural specification from embryonic stem cells. Curr Opin Neurobiol 20:37–43
4. Calhoun JD, Lambert NA, Mitalipova MM et al (2003) Differentiation of rhesus embryonic stem cells to neural progenitors and neurons. Biochem Biophys Res Commun 306:191–197
5. Elkabetz Y, Panagiotakos G, Al Shamy G et al (2008) Human ES cell-derived neural rosettes reveal a functionally distinct early neural stem cell stage. Genes Dev 22:152–165
6. Lee G, Kim H, Elkabetz Y et al (2007) Isolation and directed differentiation of neural crest stem cells derived from human embryonic stem cells. Nat Biotechnol 25:1468–1475
7. Ying QL, Stavridis M, Griffiths D, Li M, Smith A (2003) Conversion of embryonic stem cells into neuroectodermal precursors in adherent monoculture. Nat Biotechnol 21:183–186
8. Zhang SC, Wernig M, Duncan ID, Brustle O, Thomson JA (2001) In vitro differentiation of transplantable neural precursors from human embryonic stem cells. Nat Biotechnol 19: 1129–1133
9. Lazzari G, Colleoni S, Giannelli SG et al (2006) Direct derivation of neural rosettes from cloned bovine blastocysts: a model of early neurulation events and neural crest specification in vitro. Stem Cells 24:2514–2521
10. Wilson PG, Stice SS (2006) Development and differentiation of neural rosettes derived from human embryonic stem cells. Stem Cell Rev 2:67–77
11. Gardner DK, Lane M, Spitzer A, Batt PA (1994) Enhanced rates of cleavage and development for sheep zygotes cultured to the blastocyst stage in vitro in the absence of serum and somatic cells: amino acids, vitamins, and culturing embryos in groups stimulate development. Biol Reprod 50:390–400

Index

Ramiro Alberio (ed.), *Epiblast Stem Cells: Methods and Protocols*, Methods in Molecular Biology, vol. 1074,
DOI 10.1007/978-1-62703-628-3, © Springer Science+Business Media New York 2013

C

D

G

H

I

R

MIX
Papier aus verantwortungsvollen Quellen
Paper from responsible sources
FSC® C105338

If you have any concerns about our products,
you can contact us on
ProductSafety@springernature.com

In case Publisher is established outside the EU,
the EU authorized representative is:
Springer Nature Customer Service Center GmbH
Europaplatz 3, 69115 Heidelberg, Germany

Printed by Libri Plureos GmbH
in Hamburg, Germany